Medica Sacra

RICHARD MEAD

London 1740

TABLE OF CONTENTS

MEMOIRS OF THE LIFE AND WRITINGS OF THE LATE DR. MEAD

IT is a natural, nor can it be deemed an illaudable curiosity to be desirous of being informed of whatever relates to those who have eminently distinguished themselves for sagacity, parts, learning, or what else may have exalted their characters, and thereby entitled them to a degree of respect superior to the rest of their cotemporaries. The transmission of such particulars, has ever been thought no more than discharging a debt due to posterity; wherefore it is hoped, that what is here intended to be offered to the publick, relative to a gentleman, who is universally allowed to have merited so largely in the republic of letters, and more particularly in his own profession, a profession, not less useful than respectable, will not be judged impertinent or disagreeable.

Our learned author was descended from a distinguished family in Buckinghamshire, and born at Stepney the second of August 1673. His father, Mr. Matthew Mead, was held in great esteem as a divine among the presbyterians, and was possessed, during their usurped power, of the living of Stepney; from whence he was ejected the second year after the restoration of king Charles the IId. Nevertheless, tho' he had fifteen children, of whom our Richard was the seventh, he found means, with a moderate fortune, to give them a compleat education. To this purpose he kept a tutor in his house to instruct them, and they were taught latin rather by practice than by rules.

Party-rage perhaps never run higher than about the latter end of Charles the IId's reign; hereby this little domestic academy was dispersed in 1683. The king, or rather his ministers, were determined to be revenged on those, whom they could not prevail on to concur with their measures. Mr. Mead (the father) was accused of being concerned in some designs against the court; wherefore being conscious that even his being a presbyterian, rendered him obnoxious to those in power, he chose rather to consult his security by a retreat, then to rely upon his innocence; to this purpose he sought and found that repose in Holland, which was denied him in his own country; having first placed his son Richard at a school, under the tuition of an able master of his own principles: under whose care our young gentleman, by a ready genius, strong memory, and close application, made a great proficiency. At seventeen years of age he was sent to Utrecht, to be further instructed in liberal knowledge, by the celebrated Grævius, with whom he continued three years.

Having determined to devote his attention to medicine, he removed from Utrecht to Leyden, where he attended Dr. Herman's botanical lectures, and was initiated into the theory and practice of physick, by the truely eminent Dr. Pitcairn, who then held the professorial chair of this science in that university: here our young student's assiduity and discernment, so effectually recommended him to the professor, who was not very communicative of his instructions out of the college, that he established a lasting correspondence with him, and received several observations from him, which he inserted in one of his subsequent productions.

His academical studies being finished, Mr. Mead sought further accomplishments in Italy, whither he was accompanied by his elder brother, Mr. Polhill, and Dr. Thomas Pellet, afterwards president of the college of physicians.

In the course of this tour, Mr. Mead commenced doctor in philosophy and medicine at Padua, the twenty-sixth of August 1695, and afterwards spent some time at Naples and Rome: how advantageous to himself, as well as how useful to mankind he rendered his travels, his works bear ample testimony.

About the middle of the year 1696, he returned home, and settled at Stepney, in the neighbourhood where he was born: the success, he met with in his practice here, established his reputation, and was a happy presage of his future fortunes. If it be remembered, that our author was, when he began to practise, no more than twenty-three years old, that only three years, including the time taken up in his travels, were appropriated to his medical attainments, it may be, not unreasonably, admitted, that nothing but very uncommon talents, join'd to an extraordinary assiduity, could have enabled him to distinguish himself, at this early a period of life, in so extensive, and so important a science.

In 1702, Dr. Mead exhibited to the public, a manifest evidence of his capacity for, as well as application to medical researches, in his mechanical account of poisons; which he informs us was begun some years before he had leisure to publish it. These subjects, our author justly observes, had been treated hitherto very obscurely, to place therefore the surprizing phœnomena, arising from these active bodies in a more intelligible light, was his professed intention; how well he succeeded, the reception this piece universally met with, even from its first publication,[2] sufficiently declares. In 1708 he gave a new edition of it, with some few additions, the principal of which consists in some strictures on the external use of mercury in raising salivations. He has considerably further explained his sentiments upon the same head, in the edition of this work printed in 1747.

This last edition has received so many additions and alterations, as might almost entitle it to the character of a new performance.——A stiffness of opinion has been but too commonly observed, especially among writers on science; and age has been seldom found to have worn out this pertinacity: a favourite hypothesis has been defended even in opposition to the most obvious experiments, with a degree of obstinacy ever incompatible with the real interests of truth. On the contrary, our ingenious author has set before his literary successors, an example of sagacity and fortitude, truely worthy of imitation, in the victory he obtained over these self-sufficient pre-possessions; length of years was so far from rivetting in him an inflexibility of sentiment, that, joined to a most extended experience, it served only to teach him, that he had been mistaken: his candid retraction of what he thought to have been advanced amiss by himself, cannot be better expressed than in his own words. "Neither have I, says he,[3] been ashamed on some occasions, (as the Latins said) cædere vineta mea, to retrench or alter whatever I judged to be wrong. Dies diem docet. I think truth never comes so well recommended, as from one who owns his error: and it is allowed that our first master never shewed more wisdom and greatness of mind, then in confessing his mistake, in taking a fracture of a skull, for the natural suture;[4] and the compliment, which Celsus[5] makes to him on this occasion, is very remarkable and just;" nor is it less applicable to Dr. Mead at present than it was to the Coan sage in his day. "More scilicet, inquit, magnorum virorum, and fiduciam magnarum rerum habentium. Nam levia ingenia, quia nihil habent, nihil sibi detrahunt: magno ingenio, multaque nihilominus habituro, convenit etiam simplex veri erroris confessio; præcipueque in eo ministerio, quod utilitatis causâ posteris traditur."

The insertion of additions and improvements in the title of new editions of books, has been too generally, though sometimes justly, understood as little else than a contrivance of the bookseller, to animate a languishing sale; but this is far from being the case in respect to the works of our author, whose maturer sentiments on many of the subjects, he had before treated of, cannot be well comprehended, unless by a careful perusal of his later corrections, seeing the alterations he has thought fit thereby to make in his earlier productions, are not less necessary to be attended to by the prudent practitioner, than they are really interesting to the unhappy patient: the truth of which cannot be more manifestly evinced, than by his last publication of his essays on poisons; wherein he entirely subverts his former hypothesis, and builds his reasonings upon a new foundation; he also tacitly admits his former experiments to have been too precipitately made, and the conclusions deduced from them, to have been too hastily drawn.

To illustrate what has been advanced upon this head, it will not be improper to observe, that when Dr. Mead first wrote these essays, he was of opinion, "That the effect of poisons, especially those of venemous animals, might be accounted for, by their affecting the blood only: but the consideration of the suddenness of their mischief, too quick to be brought about in the course of the circulation, (for the bite of a rattle snake killed a dog in less than a quarter of an hour)[6] together with the nature of the symptoms entirely nervous, induced him to change his sentiments,[7]" and to conclude, that the poison must be conveyed by a medium of much greater quickness, which could be no other than the animal spirits.

From hence our author is led to prefix to the last edition of this performance, an inquiry into the existence and nature of this imperceptible fluid, with which we have been but very imperfectly acquainted. He has also added several new experiments, tending to confirm this theory, and explain the properties of the viperine venom, particularly by venturing to taste it; at the same time he has likewise contradicted some of those he had formerly made, whereby he had been induced to believe, this poison partook of a degree of acidity: for instance, he formerly asserted that he had seen this sanies, "as an acid, turn the blue tincture of heliotropium, to a red colour;[8]" whereas his more modern trials convinced him, it produced no alteration at all.

The essays on the tarantula and mad dog, are likewise considerably enlarged in the last impression; especially the latter, in which is now comprehended a regular and elegant history of the symptoms attending the bite of this enraged animal, the reason of the consequent hydrophobia, and more extensive directions for the cure: also an accurate description of the lichen cinereus terrestris, its efficacy, and manner of acting. A composition of equal parts of this plant and black pepper, was inserted, at our author's desire, into the London dispensatory, in the year 1721, under the title of pulvis antilyssus, which he afterwards altered by using two parts of the former, and only one of the latter, as it now stands: in 1735 he also recommended the use of this medicine in a loose sheet, intitled, a certain cure for the bite of a mad dog.

In treating of poisonous minerals, exclusive of what is added concerning mercurial unctions, our author has given a new analysis of the antient and modern arsenic; and his essay on deliterious plants, has afforded him an opportunity of enquiring into the cicuta, so much in use of old for killing, especially at Athens, and which is said to have been administered to Socrates in consequence of his condemnation. To this he has likewise subjoin'd an appendix, concerning the mischievous effects of the simple water distilled from the lauro-cerasus, or common laurel, which were first observed some years since in Ireland, where, for the sake of its flavour, it was frequently mixed with brandy.—His observations upon venemous exhalations, are not less extended, nor ought the, as well useful as ornamental, plates added to this last edition, to pass unnoticed, particularly, "The anatomical description of the parts in a viper, and in a rattlesnake, which are concerned in their poison," by our great anatomist the learned and ingenious Dr. Nichols.

In 1703 Dr. Mead communicated to the royal society, a letter published in Italy in 1687 (a copy of which he met with in the course of his travels) from Dr. Bonomo to Seignor Redi, containing some observations concerning the worms of human bodies;[9] whereby it is intended to prove, that the disease, we call the itch, proceeds merely from the biting of these animalcules: this opinion is espoused by our author in one of his latest performances,[10] wherein therefore he directs only topical applications for the cure of this troublesome disease.

The proofs our young physician had already given of literary merit, recommended him soon after the above-mentioned communication, to a seat among that learned body; in the same year he was also elected one of the physicians of St. Thomas's hospital, and was employed by the surgeons company to read anatomical lectures at their hall, which he continued to do for some years.

In 1704 appeared his treatise de imperio solis ac lunæ in corpora humana, and morbis inde oriundis. At this time the Newtonian system of philosophy, from whence our author had chiefly deduced his reasonings upon this abstruse subject, were neither thoroughly understood, nor universally received: nevertheless whatever cavils were raised against his hypothesis, it was generally admitted, that his observations had their uses in practice.

The doctor thought proper to revise this juvenile production, and to give a new edition of it in 1748; when he not only altered the disposition of some of the old, but also introduced more than a little new matter into that work: particularly he has placed some mathematical points in a clearer light, than they before appeared; he has entered into the discussion of "a difficult question, which has raised great contention among philosophers: viz. whereas water is more than eight hundred times heavier than air, how does it happen, that the latter when replete with watery vapours, depresses the mercury in the barometer; so that its fall is an indication of rain?[11]" he has also enquired into "the weight of the atmosphere on a human body, and its different pressure at different times;[12]" and he has illustrated and confirmed the medicinal part by several additional observations and cases, that promise real utility to the practice of physic. To the whole is now first adjoined a corollary tending to strengthen his reasonings upon the subject, by observations of the effects of storms on the human body; wherein, from the case of a lady who was seized in an instant with a gutta serena, (that rendered her totally blind) on the night of the great storm which happened in 1703, he is led to give a distinct account of the cause and cure of that melancholly distemper. This work is also remarkably distinguished by many curious observations our author received from his ingenious preceptor in the art of healing, Dr. Pitcairne.

Our author's distinguished genius for, and sedulous attention to the interests of his profession, procured him an acquisition of farther honours, as well as recommended him to the patronage of the most eminent of the faculty: in 1707 his Paduan diploma for doctor of physick, was confirmed by the university of Oxford; in 1716 he was elected fellow of the college of physicians, and served all the offices of that learned body, except that of president, which he declined when offered to him in 1744. Radcliff, the most followed physician of his day, in a particular manner espoused Dr. Mead, and in 1714, upon the death of the former, the latter succeeded him in his house, and the greater part of his practice; some years before which, he had quitted Stepney, and had resided in Austin Fryars.

Party-principles were far from influencing his attachments; though he was himself a zealous whig, he was equally the intimate of Garth, Arbuthnot, and Friend: his connections, more especially, with the latter, are manifested not only in their mutual writings, (of which, more hereafter) but in that when Dr. Friend was committed a prisoner to the Tower in 1723, upon a suggestion of his being concerned in the practices of Bishop Atterbury against the government, Dr. Mead became one of his securities to procure his enlargement.

In 1719, an epidemic fever made great ravages at Marseilles; and tho' the French physicians were very unwilling to admit, this disease to have been of foreign extraction or contagious; yet our government wisely thought it necessary, to consider of such measures as might be the most likely to prevent our being visited by so dangerous a neighbour; or in failure thereof, to put an early stop to the progress of the infection. Dr. Mead, whose deserved reputation may not unjustly be said to have merited that mark of distinction, was consulted on these critical and important points, by command of their excellencies, the lords justices of the kingdom, in his majesty's absence: how equal he was to this momentous talk, sufficiently appears from the discourse he published on that occasion: the approbation this performance met with, may be estimated from the reception it universally found; seven impressions were sold of it in the space of one year, and in the beginning of 1722, the author gave an eighth, to which he prefixed a long preface, particularly calculated to refute what had been advanced in France, concerning the absence of contagion in the malady that had afflicted them: he also now added a more distinct description of the plague, and its causes; and confirmed the utility of the measures he had recommended, for preventing its extension, from examples of good success, where the same had been put in practice: to these he has likewise annexed, a short chapter relating to the cure of this deplorable affliction.— In 1744, this work was carried to a ninth edition, wherein, to use the doctor's own expression, he has "here and there added some new strokes of reasoning, and, as the painters say, retouched the ornaments, and heightened the colouring of the piece." Here it may not be improper to take notice, that it is in this last impression of his discourse on the plague, that our author appears to have first adopted his theory of the properties and affections of the nervous fluid, or animal spirits, upon which he has also founded his latter reasonings on the subject of poisons, as well as in respect to the influence of the sun and moon on human bodies.

In 1723, Dr. Mead was appointed to speak the anniversary Harveian oration, before the members of the college of physicians, when, ever studious of the honour of his profession, he applied himself to wipe off the obloquy, thought to be reflected upon it, by those who maintained the practice of physic at Rome, to have been confined to slaves or freed-men, and not deemed worthy the attention of an old Roman: which oration was made publick in 1724, and to it was annexed, a dissertation upon some coins, struck by the Smyrnæans, in honour of physicians.[13]

This publication was smartly attacked by Dr. Conyers Middleton in 1726,[14] who was replied to by several, and particularly, as it is said, by Dr. John Ward, professor of rhetoric in Gresham College. This gentleman was supposed by his opponent, to have been employed by Dr. Mead, who did not chuse to enter personally, into this little-important debate; upon which presumption, Dr. Middleton published a defence of his former dissertation in the succeeding year;[15] wherein he treats his respondents with no little contempt.[16] The merits of this dispute are not intended to be here discussed, but it may not be amiss to observe, that however displeased Dr. Middleton may have been with his antagonists; in a work published several years after, he speaks of our author in the most respectful manner. In treating of an antique picture, he says, he believes it to be the first, and only one of the sort ever brought to England, "donec Meadius noster, artis medicæ decus, qui vita revera nobilis, vel principibus in republica viris, exemplum præbet, pro eo, quo omnibus fere præstat artium veterum amore, alias postea quasdam, and splendidiores, opinor, Roma quoque deportandas curavit."[17]

In respect to this controversy, our author's eulogist[18] takes notice that there is reason to believe, that Dr. Mead himself had some thoughts of more determinately explaining or confirming his sentiments upon this subject, in a work which he left unfinished, and which was designed to have been intitled, medicina vetus collectitia ex auctoribus antiquis non medicis.

However, this literary altercation, did not in the least affect our author's medical reputation, for in 1727, soon after his present Majesty's accession to the throne, whom he had the honour to serve in the same capacity while prince of Wales, he was appointed one of the royal physicians, and he had the happiness to see his two sons-in-law, Dr. Willmot and Dr. Nichols, his co-adjutors in that eminent station.

After having spent near fifty years in the constant hurry of an extensive and successful practice; after having lived (truely according to his own motto, non sibi sed toti) beyond that period assigned by the royal psalmist for the general term of mortality; when the infirmities of age would no longer permit him the free exercise of those faculties, which he had hitherto so advantageously employed in the service of the community, far from sinking into a supine indolence, or assuming a supercilious disregard of the world, he still continued his application, even in the decline of life, to the improvement of physic, and the benefit of mankind.

When he was grown unequal to the discharge of more active functions, and a retirement was become absolutely necessary, he took the opportunity of revising all his former writings: to this retreat therefore, and the happy protraction of so useful a life, the world is indebted for the improvements that appear in the latter editions of those works, which have already been taken notice of. It was not till now that our author could find leisure to perfect his discourse on the small pox and measles,[19] which had been begun by him many years before.

As it was the principal design of these memoirs, to lay before the public a concise and comprehensive history of Dr. Mead's writings, the occasion of this universally admired performance, cannot be better given than from the author's own account, contained in the preface to it, in which also his connections with, and attachment to Dr. Friend, are further illustrated.

It appears that Dr. Mead, from having observed in the year 1708, that some of his patients in St. Thomas's Hospital, recovered from a very malignant sort of the small pox, even beyond expectation, by a looseness seizing them on the ninth or tenth day of the disease, and sometimes earlier, first took the hint to try what might be done by opening the body with a gentle purge, on the decline of the distemper; finding the success of this experiment in a great measure answerable to his wishes, he communicated this method of practice to Dr. Friend, and met with his approbation.

The latter being, soon after, called to a consultation with two other eminent physicians, on the case of a young nobleman who lay dangerously ill of the small pox, proposed our author's method; this was opposed till the fourteenth day from the eruption, when the case appearing desperate, they consented to give him a gentle laxative draught; which had a very good effect: Dr. Friend was of opinion to repeat it, but was over-ruled, and the patient died the seventh day after.[20]

From the result of this case, the gentlemen of the faculty were greatly divided in opinion, as to the rectitude of this practice, insomuch that Dr. Friend thought himself under a necessity of vindicating it; and therefore sent to our author for the purport of their former conversation upon this topic, desiring it might be reduced into writing. Such was the friendship that mutually subsisted between these learned men, that this request was granted without hesitation, and Dr. Mead's letter was shewn to Dr. Radcliffe, who prevailed upon our author to consent, that the same might be annexed to Dr. Friend's intended defence; which, however he was advised by some friends, to drop at that time; whereby this letter lay by till the latter's publication of the first and third books of Hippocrates's epidemics, illustrated with nine commentaries concerning fevers. Of these the seventh treats of purging in the putrid fever, which follows upon the confluent small pox: to which are annexed, in support of this opinion, letters from four physicians on that subject, and among them that from our author, which he had translated from the english into latin, enlarged and new modelled to serve this purpose.

This work gave rise to a controversy, maintained with an unbecoming warmth on both sides: among Dr. Friend's principal opponents, may be reckoned Dr. Woodward; who, not contented with condemning a practice, experience has since evinced not only salutary in general, but in many cases absolutely necessary; likewise treated its favourers with contempt and ill-manners, and more particularly our author;[21] whose resentment upon this occasion, appears to have been carried to a justly exceptionable length, seeing it had not subsided twenty years after the death of his antagonist.[22]

Dr. Mead's daily acquisition of knowledge and experience, enabled him to enlarge to many beneficial purposes, this performance, which, in all probability, was at first designed only to illustrate and vindicate the sentiments contained in the aforementioned letter; and it is but justice to say, the applause it has found among the learned, as well for the elegance of its diction, as the perspicuity of its precepts, is no more than what is truely due to it.——To this discourse is subjoin'd a latin translation, from the arabic of Rhazes's treatise on the small pox and measles, a copy of the original having been obtained for this purpose by Dr. Mead, from the celebrated Boerhaave, between whom there had long subsisted an intimate correspondence, nor did their reciprocally differing in some opinions, diminish the friendship they mutually manifested for each other.

The year 1749, furnished two new productions from our author; a translation of one of which follows these memoirs. The other is entitled, a discourse on the scurvy, affixed to Mr. Sutton's second edition of his method for extracting the foul air out of ships.

It is more than possible that, but, for the patronage of Dr. Mead, this contrivance, which confers no less honour to the inventor, than utility to the public, might have been for ever stifled: our author, than whom no one more ardently wished for, or more zealously promoted the glory and interest of his country, being thoroughly convinced of its efficacy, so earnestly, and so effectually recommended it to the lords of the admiralty, as to prevail over the obstinate opposition that was made against its being put into practice. To the same purpose in 1742, he explained the nature and conveniencies of this invention to the royal society,[23] and with the same view he confessedly wrote the last mentioned discourse, of which he made a present to Mr. Sutton.

His medical precepts and cautions, which appeared in 1751, and was his last publication, affords an indisputable testimony, that length of years had not in the least impaired his intellectual faculties. Our author has herein furnished the public, with the principal helps against most diseases which he had either learned by long practice, or deduced from rational principles.[24] Who could with the same propriety take upon himself to be an instructor and legislator in the medical world, as he who had been taught to distinguish truth from falsehood, in the course of so extended an experience, protracted now to almost threescore years? to this may be added, that he has so contrived to blend the utile dulci, by embellishing his precepts with all the delicacy of polite expression, as to render them at the same time not less entertaining than instructive.

However, this work was productive of two other little pieces, from two gentlemen of the faculty: one by Dr. Summers; who in a pamphlet on the success of warm bathing in paralytic cases, controverts Dr. Mead's assertion, that "hot bathing is prejudicial to all paralytics" ... "calidæ vero immersiones omnibus paralyticis nocent[25]."—Some reflections upon the advocates for Mrs. Stephens's medicines, in the cure of the stone and gravel, by our author, occasioned a letter to him on that subject by Dr. Hartley of Bath. The former expressed himself in the following manner; "Neque temperare mihi possum, quin dicam in opprobrium nuper medicis nonnullis cessisse, quod insano pretio redimendi anile remedium magnatibus auctores fuerunt.[26]" ... "Nor can I forbear observing, tho' I am extremely sorry for the occasion, that some gentlemen of the faculty a few years since acted a part much beneath their characters, first in suffering themselves to be imposed on, and then in encouraging the legislature to purchase an old woman's medicine at an exorbitant price."[27] Of this the latter complains as an unmerited indignity, "Illud interea (inquit) tanquem inopinatum, and ab æquitate tua alienum queri liceat, Te, qui in obvios quoscunque comis and urbanus esse, bene autem merentibus de re medica, vel etiam literaria quavis, summa cum benignitate favere soleas, in lithrontriptici fautores acerbiùs invectum fuisse; and non potius laudi illis dedisse, quod arcanum sine pretio vulgatum, virorum dignitate, fide, ingenio, artis nostræ peritiâ illustrium examini subjecerent, neque aliam viam ad præmium reportandum aperiri voluerint, quam quæ, veris licèt rerum inventoribus facilis and munita, jactatoribus tamen and falsiloquis esset impervia.[28]" ... In the mean while, I cannot but complain of it as a thing unexpected, and greatly inconsistent with your usual candour, that YOU, who are so courteous and humane to all mankind, and so remarkably the patron of those who excel in the profession of physic, or indeed in any branch of learning, should so severely reproach the favourers of this lithontriptic medicine; and not rather have commended them, for submitting a secret, communicated to them without fee or reward, to the examination of some worthy physicians, eminent for integrity, ingenuity, and learning: and for endeavouring to excite the munificence of the publick in such a manner only, as to render it accessible to the true authors of an important discovery, but impervious to boasting impostors.

In enumerating the obligations the republic of letters is under to Dr. Mead, it would be injustice to omit taking notice, that to his generosity and public spirit, it is farther indebted for the first complete edition of the celebrated history of Thuanus.[29]

To enlarge upon his literary collections, and other curiosities, would at present be useless, seeing the world will soon be apprized of their value and contents from the catalogues that are already, and are yet about to be published of them; it may therefore suffice to say, that he did not shew more assiduity and judgment in collecting them, than he did candour and generosity in permitting the use of them to all that were competent judges, or that could benefit themselves, or the public by them.

It may, perhaps not unjustly, be said no Subject in Europe had a cabinet so richly and so judiciously filled; to which the correspondence he maintained with the learned in all parts of Europe, not a little contributed; nor can there be an higher instance given of his reputation in this respect, than in the king of Naples having sent him the two first volumes of M. Bajurdi's account of the antiquities found in Herculaneum, with the additional compliment of asking in return, only, a compleat collection of our author's works, to which was adjoined, an invitation to visit that newly discovered subterraneous city: an invitation that could not but be greatly pleasing to a genius so inquisitive after knowledge, and which he declared, he should very gladly have embraced, had not his advanced years been an insuperable impediment, to the gratification of his curiosity. In short, his character abroad was so well known and established, that a foreigner of any taste, would have thought it a reproach to him, to have been in England without seeing Dr. Mead.

As his knowledge was not limited only to his profession, the deserving in all sciences had not only free access to him, but always found a welcome reception, and at his table might daily be seen together the naturalist, the antiquarian, the mathematician, and the mechanic, with all whom he was capable of conversing in their respective terms; here might be seen united the magnificence of a prince, with the pleasures of the wise.

His munificence was conspicuous in that there was no remarkable publick charity to which he was not a benefactor, particularly he was one of the earliest promoters of, and subscribers to the Foundling hospital.

Let these specimens of his superior abilities and merit suffice for the present, nor let envy or detraction attempt to sully so exalted a character.— Soon after the publication of his monita and præcepta medica, this ornament of his profession, and delight of his acquaintance, grew more and more sensible of the natural infirmities attending his length of years; and with the utmost tranquillity and resignation, quietly sunk into the arms of death on the 16th of February 1754. To whom may, with the greatest propriety, be applied a part of the epitaph inscribed to the memory of the celebrated Guicciardini, at Florence;

Cujus Otium an Negotium
Gloriosius incertum:
Nisi Otii Lumen Negotii Famam
Clariorem reddidisset.

THE END.

FOOTNOTES:

[1] Mr. Nathaniel Mead, who was at first destined to the service of religion, and preach'd two or three times at the meeting house at Stepney, built by his father, after his ejection from the parish church: but taking a dislike to theological studies, he applied himself to the law, and made as great a figure at the bar, as his brother did in physick.

[2] An abstract of this work was thought deserving a place in the philosophical transactions (N° 283) for the months of January and February 1703.

[3] Advertisement prefixed to the last edition of the essay on poisons, p. 4.

[4] Epidem. lib. iv. § 14.

[5] Medicin. lib. viii. c. 4.

[6] Philosophical transactions N° 399.

[7] Introduction to the last edition of the essays on poisons, page 12.

[8] Second edition of those essays, page 10.

[9] An abstract of part of this letter was inserted in the before-cited number of the philosophical transactions. Vid. supra p. 10.

[10] Monita and præcepta medica, p. 211, andc.

[11] Stack's translation of the influence of the sun and moon, p. 21.

[12] Ibid. p. 30.

[13] Dissertatio de nummis quibusdam, a Smyrnæis, in medicorum honorem, percussis.

[14] In a piece entitled, De medicorum apud veteres Romanos degentium conditione dissertatio; contra viros celeberrimos Jac. Sponium and Rich. Meadium, M.D.D. Servilem atque ignobilem eam fuisse ostenditur, published in the fourth volume of his works, p. 179.

[15] Dissertationis, andc. contra anonymos quosdam notarum brevium, responsionis atque animadversionis auctores, desensio, ibid. p. 207.

[16] Speaking of the answer ascribed to Dr. Ward, Dr. Middleton says, quamvis enim nomen suum celavisset, sensi tamen hominem e rhetorum turba conductum esse oportere; cui scilicet generi concessum novimus, omnia tragice ornare, augere, ementiri: is mihi solum scrupulus restabat, quod in ejus quidem sermone, nihil plane, quod rhetorem oleret, nihil venustatis, nihil ornatùs, sed inculta potiùs omnia nec satis latina invenirem. Hujusmodi itaque scriptorem, haud magis quam alterum illum (cui neutiquam sane eum anteserendam censeo) cogitatione ulla mea aut animadversione dignum judicassem; ni hanc potissimum hominem a clarissimo Meadio ad hoc respondendi munus delectum; librumque ipsum ejusdem cura and sumptibus in lucem emissam; amicisque suis manu propria inscriptum and dono a Meadio ipso missum intellixissem.

[17] Germana quædam antiquitatis erudita monumenta, andc. first published in 1745, and inserted in the before-cited volume of his works, p. 2.

[18] The ingenious Dr. Maty, who in his journal britannique (a work not less useful than entertaining) for the months of July and August 1754, has inserted a piece, which he titles, eloge du docteur Richard Mead, composed, as himself takes notice, from materials communicated to him by Mr. Birch; to which piece these memoirs are obliged for some anecdotes relating to our learned author.

[19] De variolis and morbillis 1747.

[20] Friendi opera, p. 263.

[21] The state of physic, by John Woodward, M.D. printed in 1718.

[22] "In the front of this band stood forth Dr. John Woodward, physic professor at Gresham College, a man equally ill-bred, vain, and ill-natured; who, after being for some time apprentice to a linnen-draper, took it into his head to make a collection of shells and fossils, in order to pass upon the world for a philosopher; thence getting admission into a physician's family, at length, by dint of interest, obtained a doctor's degree." Preface to the discourse on the small pox, andc. p. 8, andc.

[23] In a paper read before the royal society, Feb. 11, 1741-2, and published

in Mr. Sutton's account, page 41. He also presented a model of this invention made in copper to the royal society, which cost him 200l.

[24] Preface to the monita and præcepta medica, p. 1.

[25] Monita and præcepta, p. 62, and Stack's translation of the same, p. 69.

[26] Our author's disapprobation of this medicine and its favourers, is no less severely express in his treatise concerning the influence of the sun and moon upon human bodies, p. 100.

[27] Monita, andc. medica, and Stack's translation, p. 174 and 197.

[28] Ad virum clarissimum Ric. Mead, M.D. Epistolæ, varias lithontripticum, Joannæ Stephens exhibendi methodos indicans. Auctore Davide Hartley, A.M. p. 3.

[29] Published in seven volumes folio 1733, by Samuel Buckley, under the sanction of an act of parliament.

THE PREFACE

MY declining years having in a great measure released me from those medical fatigues, in which, for the publick good, (at least as I hope) I have been employed about fifty years, I have determined to pass the short remains of life in such a sort of leisure, as may prove neither disagreeable to myself, nor useless to others. For good men are of opinion, that we must give an account even of our idle hours, and therefore thought it necessary, that they should be always well-spent.

Having from my earliest childhood entertained a strong passion for learning, after I had chosen the art of medicine for my profession, I still never intermitted my literary studies; to which I had recourse from time to time, as to refreshments strengthening me in my daily labours, and charming my cares. Thus, among other subjects, I frequently read the holy scriptures, as becomes a christian; and next to those things which regard eternal life, and the doctrine of morality, I usually gave particular attention to the histories of diseases, and the various ailments therein recorded; comparing those with what I had learnt either from medical writers or my own experience. And this I did the more willingly, because I had remarked that divines, thro' an unacquaintance with medicinal knowledge, frequently differed widely in their sentiments; especially on the subject of dæmoniacs cured by the power of our saviour Jesus Christ. For it is the opinion of many, that these were really possessed with devils, and that his divine virtue shone forth in nothing more conspicuous than in expelling them. I am very far from having the least intention to undermine the foundations of the christian doctrine, or to endeavour, by a perverse interpretation of the sacred oracles, to despoil the Son of God of his divinity, which he has demonstrated by so many and great works performed contrary to the laws of nature. Truth stands no more in need of the patronage of error, than does a natural good complexion of paint. And it is certain, that the opinion which has been prevalent for many ages, of the power granted to devils, of torturing human bodies and minds, has been several ways made subservient to the subtle designs of crafty men, to the very great detriment and shame of the christian religion.

What sensible man can avoid justly deriding those solemn ceremonies, practised by the roman priests, in exorcising, as they are fond of terming it, dæmoniacs: while proper persons (hired and) taught to counterfeit certain gestures and fits of fury, such as are believed to be caused by evil spirits, pretend that they are freed from devils, and restored to their senses by holy water, and certain prayers, as by inchantment. But these juggling tricks, how grosly soever they may impose on the eyes and minds of the ignorant multitude, not only scandalize, but also do a real injury to, men of greater penetration. For such, seeing into the cheat, often rush headlong into impiety; and viewing all sacred things in the same light, after they have learnt

Relligionibus atque minis obsistere vatum:[30]

they advance farther, and by an abominable effort, endeavour thoroughly to root out of their minds all sense and fear of the supreme deity. In which proceeding they act as if a person doubted of the existence of the Indies, because travellers relate many falshoods and fictions concerning them. Hence it comes to pass, that, in countries too much given up to superstition, very many atheists are to be met with even among the learned, whom their learning and knowledge ought to secure from these errors. Therefore to be free from this folly, is the principal part of wisdom; next to which, is not to corrupt truth with fictitious opinions.

And indeed it is frequently to me a matter of wonder, why our spiritual guides so strenuously insist on exhibiting devils on the stage, in order to make the divinity of Christ triumph over these infernal enemies. Is Christ's divine power less manifested by the cure of the most grievous diseases, performed in an instant at his command; than by the expulsion of evil spirits out of the bodies of men? Certainly all the wonderful things done by him for the good of mankind, such as restoring sight to the blind, firmness and flexibility to relaxed or contracted nerves, calling the dead to life, changing the properties of the elements, and others of the same kind, are testimonies of the omnipotence of the creator of the world, and demonstrate the presence of God; who alone commands all nature, and at his pleasure changes and inverts the order of things established by himself. Wherefore it cannot be doubted, that He, who has perform'd these things, had the devils subject to him, that they might not obstruct his gracious resolution of revealing the will of his father to men, and correcting their depraved morals.

But to resume the subject of dæmoniacs, the opinion, which I propose in this treatise, is not purely my own, but also of several other persons, before me, eminent for piety and learning. And indeed among our own countrymen, it was in the last century defended in an excellent dissertation, by that treasure of sacred knowledge, the reverend Joseph Mead. Wherefore as I have the honour to be of the same family with him, and am the son of Matthew Mead, a very able divine, I always thought I might lay some claim to these studies, by a kind of hereditary right.

I am not insensible of the difficulty of removing vulgar errors, especially those which relate to religion. For every body knows the power of education, in imprinting on the mind notions, which are hard to be effaced even in adult age. Children in the dark, fear ghosts and hobgoblins; and hence often quake with the same fear through the whole course of their lives. Why then do we admire, if we can hardly unlearn, and clear our minds of, some false notions, even when we are advancing to old age? Nor will this be deemed indeed a matter of little importance by him, who considers the serious evils, into which mankind are often led, by things that to some may appear trifling, as being nothing more than bugbears of children and women. My soul is seized with horror on recollecting, how many millions of innocent persons have been condemned to the flames in various nations, since the birth of Christ, upon the bare suspicion of witchcraft: while the very judges were perhaps either blinded by vain prejudices, or dreaded the incensed populace, if they acquitted those, whom the mob had previously adjudged guilty. Who would believe that any man in his right senses could boast, as a[xiv] matter of merit, that he had capitally condemned about nine hundred persons for witchcraft, in the space of fifteen years, in the sole dutchy of Lorraine?[31] And yet from many histories, which he relates of those who suffered, it manifestly appears, that every individual of these criminals, had no compacts with devils, as they themselves imagined, but were really mad, so as openly to confess that they had done such feats as are impossible in the nature of things. But so it happens, that error generally begets superstition, and superstition cruelty. Wherefore I most heartily rejoice, that I have lived to see all our laws relating to witchcraft entirely abolished: whereas foreign states still retain this barbarous cruelty, and with various degrees of obstinacy in proportion to their ignorance of natural causes. And it is but too true, that the doctrine of dæmons is so understood by the vulgar, as if the devil was to be esteemed a sort of deity; or at least, that, laying the fear of him aside, no divine worship can well subsist; altho' the apostle has expresly said; For this purpose the Son of God was manifested, that he might destroy the works of the devil.[32]

And here it may not be improper, once for all, to inform the reader, that I have generally made use of Sebastian Castalio's version of the bible, because, upon collating it in many places, I found it to be not only excellent Latin, but also very accurate, and particularly well adapted to the sense and meaning of the words in the Hebrew and Greek.

Nor can I refrain from declaring, that I have not writ these essays for the profane or vulgar; but for those only who are well versed, or at least initiated in theological or medical studies: and for this reason I chose to publish it in Latin; which language has for many ages past been made use of by learned men; in order to communicate to each other, whatsoever might seem to them either new, or expressed in a different manner from the common notions. Wherefore if any person should intend to publish an English version of this book, I give him this timely notice, that he will do it, not only against my will; but likewise in direct opposition to that equitable law, whereby every man is allowed to dispose of his own property according to his pleasure.[33]

But to bring this preface to a conclusion; it is manifest that the christian religion requires of all its members in a most especial manner, to practice every act of humanity and benevolence towards each other. Wherefore the utmost care ought to be taken, that this beneficent disposition of mind be not corrupted by any means whatsoever: and nothing contributes more towards bringing on this corruption, than opinions derogatory from the divine goodness. Upon this account, as such is the misfortune of our times, that it is not only allowed, but even by many deemed a commendable action, to oppugn, and by every method to invalidate, the doctrine and authority of the christian religion; no interpretations of the histories of miracles ought to be look'd upon as out of season, provided they appear neither improbable, nor repugnant to the nature of the facts related.

In fine, it was not my intention to treat of every disease mentioned in holy writ; but to confine myself more particularly to those, the nature of which is generally but little known, or at least to such as I had some peculiar medicine for, or method of cure, to offer to the public; and to perform this task, in the same order, in which they occur in those sacred writings: excepting only Job's disease, to which I have given the first place, on account of the great antiquity of that book. The Saviour of the world, in order to make his divine power manifest to mankind, cured many other diseases, both of the body and mind, besides those which I have mentioned in this work: the nature and causes of all which diseases, whosoever would intend to enquire into, must of necessity compile a body of physic, which was not my present design. But if providence protract my life, I am not without hopes of laying more of my thoughts on this subject before the public, for the honour which I bear to my profession, unless

Frigidus obstiterit circum præcordia sanguis.

In the mean time, whatever be the fate of these essays with my readers, I shall rest satisfied from a consciousness of the rectitude of my intention, in having thus employ'd some of my hours of leisure.

FOOTNOTES:

[30] Lucret. Lib. i. ver. 110.
[31] See Nic. Remigii Dæmonolatreia.
[32] John. Ep. i. Chap. iii. ver. 8.
[33] This declaration seems to have been intended only to prevent any surreptitious translation of this performance from appearing, seeing most of the works of our learned author have heretofore been greatly disgraced by attempts of that kind. Nevertheless the public may be assured, that Dr. Mead not only approved, but inspected what is now offered to them.

A COMMENTARY ON THE DISEASES MENTIONED IN SCRIPTURE

THE DISEASE OF JOB

Job's disease is rendered remarkable by some uncommon circumstances and consequences; such as the dignity of the man, the sudden change of his condition, his extraordinary adversity, his incredible patience under them, his restoration to a much happier state than he had ever before enjoyed, and lastly the singular nature of the illness with which he was seized.

His habitation was in the land of Uz, which, according to the learned Friderick Spanheim,[34] was situated in the northern part of Arabia deserta, towards the Euphrates and Mesopotamia. He was a very illustrious man, the most opulent of all the Orientals, very happy in sons and daughters, of a most upright life and exemplary piety. Now it is related that God, in order to try his integrity and constancy, permitted Satan to afflict him by all means which he could devise, except the taking away of his life. "In pursuance of this permission, Satan brought the most dreadful calamities on him; for all his oxen and asses were driven away by the Sabeans; his sheep and servants were consumed by fire from heaven; his camels were carried off; his sons and daughters were crush'd to death by the falling in of the house upon them in a violent storm of wind; and soon after he himself was afflicted with scabs and foul ulcers all over his body; so that he sate down among the ashes, and scraped himself with a potsherd." Thus from a very rich man he became extremely poor, and from the heighth of prosperity he sunk into the depth of misery. And yet all these evils did not give the least shock to his firmness of mind, nor to his piety towards God:[35] wherefore the Lord, moved by his prayers, put an end to all his calamities; gave him twice as much wealth as he had lost, and made him more prosperous than he had ever been before.[36]

Now the book of Job may justly be esteemed the most ancient of all books, of which we have any certain account: for some are of opinion that it was written in the times of the patriarchs; many others, that it was composed about the days of Moses, and even by Moses himself; and there are but few who think it posterior to him.[37] For my part, I embrace the learned Lightfoot's opinion, that it was composed by Elihu, one of Job's companions, chiefly because he therein speaks of himself as of the writer of this history,[38] and if so, it will appear to be older than the days of Moses. However this be, it is most certain that this book carries with it manifest tokens of very great antiquity; the most material of which seem to be these. In it there is not the least mention made of the departure of the Israelites out of Egypt, of Moses, or the Mosaic Law. After the manner of the Patriarchs, Job, as the head of his family, offered sacrifices in his own private house, for the sins of his children.[39] When he declares his integrity he scarcely mentions any other Idolatry, but that most ancient one, the worship of the sun and moon,[40] which we know to be very old, and to have first obtained among the neighbouring Chaldeans, and Phœnicians. In fine his own age, protracted far beyond the life of man in Moses's time, is a proof of its antiquity, for he lived a hundred and forty years after an end had been put to his calamities; so that it is reasonable to believe that he lived above two hundred years in all. For that he was aged, when his misfortunes crowded on him, may be hence inferred, that, altho' his three friends are stiled old men,[41] yet in his disputes with them, he does not seem to honour them for their age, as Elihu does. To avoid prolixity, I join with Spanheim in opinion, that Job's time coincides with the bondage of the children of Israel in Egypt, so as to be neither posterior to their quitting that country, nor anterior to their entering it.

But there subsists a dispute of a different nature between very grave authors, and that is, whether this narrative be a fable or a true history: If I were allowed to interpose my opinion, I would say, that it is not a parable invented by Ὑποτύπωσις, but a dramatic poem composed upon a true history; and perhaps with this design, that from the example of this illustrious and upright, yet afflicted and most miserable man, the people of Israel might learn to bear with patience, all those evils and hardships, which they were daily suffering in their Egyptian captivity. That this book is metrical, as well as David's Psalms, the Proverbs, Ecclesiastes, and Solomon's Song, is generally allowed: and the persons of the drama are God, Satan, Job and his wife, his three friends, and Elihu. Wherefore it is, says Grotius, a real fact, but poetically handled.[42] Poetry was certainly a very ancient manner of writing, and poets were wont to embellish true histories in their own way, as we see in the most ancient among the Greeks and Romans. And among the Hebrews likewise, long after the time above-mentioned, Ezechiel comprised the history of the departure out of Egypt in a dramatic poem; upon which account he is called by Clemens

Alexandrinus, the poet of Judaic tragedies.[43] Nor indeed, in my opinion, can there be found, in this kind of writing, any thing more admirable, and better adapted to move the passions than this piece; whether we regard the sublimity and elegance of style, the description of natural things, or in fine, the propriety of the characters ascribed to all the persons concerned in it; all which circumstances are of the greatest moment in a dramatic performance.

... Quo propius stes

Te capiet magis.

The nearer you behold,

The more it strikes you.

Before I close this chapter, it may not be improper to offer my conjecture concerning the disease of this illustrious man. But previous to this, it is proper to remark, that it is not Job himself, or his friends, but the author of the book that attributes his calamities to Satan; for this author's intention seems to be, to shew, by a striking example, that the world is governed by the providence of Almighty God, and as the holy angels, whose ministry God makes use of in distributing his bountiful gifts, punctually execute all his commands; so Satan himself with his agents are under the power of God, and cannot inflict any evils on mankind without the divine permission. Thus, when the Sons of God (angels) came and presented themselves before the Lord, it is said that Satan came also among them. Now the word astare to present one's self, as Moses Maimonides[44] observes, signifies to be prepared to receive Jehovah's commands, but Satan came of his own accord and mixed with them without any summons.

Now as to the disease, it is plain that it was cuticular, and as it is certain that the bodies of the Hebrews were very liable to foul ulcers of the skin from time immemorial; upon which account it is, that learned men are of opinion that they were forbid the eating of swine's flesh (which, as it affords a gross nourishment, and not easily perspirable, is very improper food in such constitutions) wherefore by how much hotter the countries were which they inhabited, such as are the desarts of Arabia, the more severely these disorders raged. And authors of other nations, who despised and envied the Jews, say that it was upon this account that they were driven out of Egypt; lest the leprosy, a disease common among them, should spread over the country.[45] But there is another much worse disease, so frequent in Egypt, that it is said to be endemial there,[46] though it may also be engendered in this hot country, I mean the Elephantiasis. Perhaps it was this, which is nearly of the same nature with the leprosy, that had affected the body of our righteous man: but on this subject we shall treat more largely in the subsequent chapter.

FOOTNOTES:

[34] Histor. Jobi, Cap. iv.

[35] See Job Chap. i and ii.

[36] The same, Chap. xiii.

[37] See Spanheim's learned dissertation on this subject in the book above quoted, Chap. viii. and ix.

[38] His Works, tom. 1. page 24.

[39] Job, Chap. i. v. 5.

[40] The same, Chap. xxxi. v. 26, 27.

[41] Job Chap. xxxii. v. 6.

[42] Est ergo res vere gesta, sed poeticè tractata. In locum.

[43] Ὁ τῶν Ἰ☐ δαἱκῶν τραγωδιῶν ποιητής. Stromat. book 1. p. 414 of the Oxford Edit. 1715.

[44] More Nevochim, Part. iii. Chap. xxii.

[45] Justin. Hist. Lib. xxxvi, C. 2. and Tacit. Hist. Lib. v. ab initio.

[46] Lucret. Lib. vi. v. 1112.

Est Elephas morbus, qui propter flumina Nisi.

Gignitur Aegypto in media.

THE LEPROSY

A most severe disease, to which the bodies of the Jews were very subject, was the Leprosy. Its signs recorded in the holy scriptures are chiefly these. Pimples arose in the skin; the hair was turned white; the plague (or sore) in sight was deeper than the skin, when the disease had been of long standing; a white tumour appeared in the skin, in which there was quick flesh; the foul eruptions gained ground daily, and at length covered the whole surface of the body. And the evil is said to infect, not only the human body, but also the cloaths and garments, nay (what may seem strange) utensils made of skins or furs, and even the very walls of the houses. Wherefore there are precepts laid down for cleansing these also, as well as the lepers.

Medical authors are of different opinions concerning the contagion of this disease. And whereas neither the Arabian nor Greek physicians, who have treated largely of the leprosy, have given the least hint of this extraordinary force of it, whereby it may infect cloaths and walls of houses; the Rabbin doctors dispute, whether that which seized the Jews, was not intirely different from the common leprosy; and they all affirm, that there never appeared in the World, a leprosy of cloaths and houses, except only in Judea, and among the sole people of Israel.

For my part, I shall now freely propose, what I think most probable on the subject. One kind of contagion is more subtile than another; for there is a sort, which is taken into the body by the very breath; such as I have elsewhere said to exist in the plague, small pox, and other malignant fevers. But there is another sort, which infects by contact alone; either internal, as the venom of the venereal disease; or external, as that of the itch, which is conveyed into the body by rubbing against cloaths, whether woollen or linnen. Wherefore the leprosy, which is a species of the itch, may pass into a sound man in this last manner; perhaps also by cohabitation; as Fracastorius has observed, that a consumption is contagious, and is contracted by living with a phthisical person, by the gliding of the corrupted and putrefied juices of the sick into the lungs of the sound man.[47] And Aretæus is of the same opinion with regard to the Elephantiasis, a disease nearly allied to the Leprosy: for he gives this caution, "That it is not less dangerous to converse and live with persons affected with this distemper, than with those infected with the plague; because the contagion is communicated by the inspired[48] air."

But here occurs a considerable difficulty. For Moses says, "If in the leprosy there be observed a white tumour in the skin, and it have turned the hair white in it, and there be quick flesh within the tumour; it is an old leprosy in the skin of his flesh. But if the leprosy spread broad in the skin, and cover the whole skin of the diseased from his head even to his feet, the person shall be pronounced[49]

clean." But the difficulty contained

in this passage will vanish, if we suppose, as it manifestly appears to me, that it points out two different species of the disease; the one in which the eroded skin was ulcerated, so that the quick flesh appeared underneath; the other, which spread on the surface of the skin only in the form of rough scales. And from this difference it happened, that the former species was, and the latter was not, contagious. For these scales, being dry and light like bran, do not penetrate into the skin; whereas the purulent matter issuing from the ulcers infects the surface of the body. But concerning the differences of cuticular diseases, I heartily recommend to the reader's perusal, what Johannes Manardus, equally valuable for his medical knowledge and the purity of his Latin, has written upon the subject.[50]

There is no time, in which this disease was not known; but it was always more severe in Syria and Egypt, as they are hotter countries, than in Greece and other parts of Europe; and it is even at this day frequent in those regions. For I have been assured by travellers, that there are two hospitals for the leprous alone in Damascus. And there is a fountain at Edessa, in which great numbers of people affected with this cuticular foulness wash daily, as was the ancient custom.

Moreover we read the principal signs, which occur in the description of the Mosaic leprosy, excepting only the infection of the cloaths and houses (of which by and by) recorded by the Greek Physicians. Hippocrates himself calls the λεύκη or white leprosy Φοινικίη ν□σος the Phœnician disease.[51]

For that the word φθινικὴ ought to be read Φοινικίη], appears manifestly from Galen in his Explicatio linguarum Hippocratis; where he says that Φοινικίη ν□σος is a disease which is frequent in Phœnicia and other eastern regions.[52] In the foregoing chapter I said that the Leprosy (Leuce) and the

Elephantiasis, were diseases of great affinity:[53] in confirmation of which notion the same Galen observes, that the one sometimes changes into the other.[54] Now these two distempers are no where better described than by Celsus, who lived about the time of Augustus Cæsar, and having collected the works of the principal Greek writers in physic and surgery, digested them into order, and turned them into elegant Latin with great judgment. Thus he describes the leprous diseases. Three are three species of the Vitiligo. It is named ἄλφος, when it is of a white colour, with some degree of roughness, and is not continuous, but appears as if some little drops were dispersed here and there; sometimes it spreads wider, but with certain intermissions or discontinuities. The μέλας differs from this in colour, because it is black, and like a shadow, but in other circumstances they agree.

The λεύκη has some similitude with the ἄλφος, but it has more of the white, and runs in deeper: and in it the hairs are white, and like down. All these spread themselves, but in some persons quicker, in others slower. The Alphos and Melas come on, and go off some people at different times; but the Leuce does not easily quit the patient, whom it has seized.[55] But in the Elephantiasis, says the same author, the whole body is so affected, that the very bones may be said to be injured. The surface of the body has a number of spots and tumors on it; and their redness is by degrees changed into a dusky or blackish colour. The surface of the skin is unequally thick and thin, hard and soft; and is scaley and rough: the body is emaciated; the mouth, legs and feet swell. When the disease is inveterate, the nails on the fingers and toes are hidden by the swelling.[56] And the accounts left us by the Arabian physicians, agree with these descriptions. Avicenna, the chief of them, says that the Leprosy is a sort of universal cancer of the whole body.[57] Wherefore it plainly appears from all that has been said, that the Syrian Leprosy did not differ in nature, but in degree only, from the Grecian, which was there called λεύκη; and that this same disease had an affinity with the Elephantiasis, sometimes among the Greeks, but very much among the Arabs. For the climate and manner of living, very much aggravates all cuticular diseases.

Now with regard to the infection of the cloaths, it has been found by most certain experiments, not only in the plague, and some other malignant eruptive fevers, as the small pox and measles, but even in the common itch; that the infection, once received into all sorts of furs or skins, woollen, linnen, and silk, remains a long time in them, and thence passes into human bodies. Wherefore it is easy to conceive, that the leprous miasmata might pass from such materials into the bodies of those, who either wore or handled them, and, like seeds sown, produce the disease peculiar to them. For it is well known, that the surface of the body, let it appear ever so soft and smooth, is not only full of pores, but also of little furrows, and therefore is a proper nest for receiving and cherishing the minute, but very active, particles exhaling from infected bodies. But I have treated this subject in a more extensive manner in my Discourse on the Plague.[58] And these seeds of contagion are soon mixed with an acrid and salt humor, derived from the blood; which as it naturally ought, partly to have turned into nutriment, and partly to have perspired through the skin, it now lodges, and corrodes the little scales of the cuticle; and these becoming dry and white, sometimes even as white as snow, are separated from the skin, and fall off like bran. Now, altho' this disease is very uncommon in our colder climate; yet I have seen one remarkable case of it, in a countryman, whose whole body was so miserably seized by it, that his skin was shining as if covered with snow: and as the furfuraceous scales were daily rubbed off, the flesh appeared quick or raw underneath. This wretch had constantly lived in a swampy place, and was obliged to support himself with bad diet and foul water.

But it is much more difficult to account for the infection of the houses. For it seems hardly possible in nature, that the leprous spots should grow and spread on dry walls, made of solid materials. But upon a serious consideration of the different substances employed in building the walls of houses, such as stones, lime, bituminous earth, hair of animals, and other such things mix'd together; I thought it probable, that they may by a kind of fermentation, produce those hollow greenish or reddish strokes in sight lower than the wall (or within the surface)[59] which, as they in some measure resembled the leprous scabs on the human body, were named the Leprosy in a house. For bodies of different natures, very easily effervesce upon being blended together. Wherefore we may reasonably suppose that this moisture or mouldiness, gradually coming forth and spreading on the walls, might prove very prejudicial to the inhabitants, by its stinking and unwholesome smell, without having recourse to any contagious quality in it. And somewhat analogous to this is pretty frequently observable in our own houses; where, when the walls are plaistered with bad mortar, the calcarious and nitrose salts sweat out upon their surface, of a colour almost as white as snow. The power of inspecting their houses was invested in the priests; who, when they observed this foulness, gave orders first to have the walls of the house scraped all around; and afterwards, if it continued to break out, to pull down the house, and carry the materials out of the city into an unclean place.

I am well aware, that all this is related, as if God himself had struck the house with this plague. But it is well known, that that way of speaking is not uncommon in the jewish history; as in unexpected evils and dreadful calamities, which are sometimes said to be done by the hand of God, tho' they may be produced by natural causes. Nor can I be easily induced to believe, with some divines, that God, who commanded his people to be always free from every sort of uncleanness, would vouchsafe to work a miracle, in order to inflict this most filthy punishment on any person. Thus much is indubitable, that the precepts of the mosaic law were constituted particularly, to avert the people from idolatry and false religion, and at the same time to keep them clear of all uncleanness.[60] To this end conspired the prohibition of eating blood, carrion, or animals that died spontaneously, swines flesh, and that of several other creatures.[61] For all these meats yield a gross nutriment, which is improper and prejudicial in diseases of the skin.

But in order to close these theological researches with somewhat medical, I am convinced from experience, that there is not a better medicine known against this filthy disease, than the tincture of Cantharides of the London Dispensatory. Its remarkable virtue in this case, is owing to the diuretic quality of these flies. For there is a great harmony between the kidneys and glands of the skin, so that the humors brought on the latter, easily find a way thro' the former, and are carried off by urine: and on the other hand, when the kidneys have failed in the performance of their functions, an urinous humor sometimes perspires thro' the cuticular pores. But such cathartics are to be interposed at proper intervals, as are most proper for evacuating thick and acrid humors.

FOOTNOTES:

[47] De morbis contagiosis. Lib. ii. Cap. ix.
[48] De causis diuturnorum morborum, et de curationibus eorundem, Lib. ii. Cap. xiii.
[49] Levit. Chap. xiii. v. 10 andc.
[50] Epist. Medicinal. Lib. vii. Epist. ii.
[51] Prorrhetic. Lib. ii. sub finem.

[52] Ἡ κατὰ Φοινίκην, κι κατὰ τὰ ἄλλα ἀνατολικὰ μέρη πλεονάζ☐ σα.
[53] Pag. 15.
[54] De simpl. medicam. facult. Lib. xi.
[55] De medicina, Lib. v. Cap. xxviii. §. 19.
[56] Lib. iii. Cap. xxv.
[57] Canon, Lib. iv. Fen. 3. Tract. 3. Cap. i.
[58] Chap. i.
[59] Levit. Chap. xiv. v. 37.
[60] Mos. Maimonid. More Navochim, Part. iii. Cap. xxxiii. et xlviii.
[61] Levit. Chap. xi. et xvii.

RICHARD MEAD

THE DISEASE OF KING SAUL

When "King Saul was abandoned by the Spirit of God, and an evil spirit from the Lord troubled him; his courtiers persuaded him to command his servants to seek out somebody that was a good player on the harp, who might sooth or compose him by his music, when the evil spirit from God was upon him." Which when Saul had done, by sending messengers for David; "whenever it happened that Saul was seized with that evil spirit, David took his harp, and play'd on it; and thus Saul was refreshed and became composed, and the evil spirit departed from[62] him."

Now to me it appears manifest, that this king's disease was a true madness, and of the melancholic or atrabilarious kind, as the ancient physicians called it. And the fits return'd on him at uncertain periods, as is frequently the case in this sort of disease. Nor could the cause of that disorder be a secret, seeing he had been lately deprived of his kingdom by God's express command. Likewise the remedy applied, to wit, playing on the harp, was an extremely proper one. For physicians have long since taught us, that symphonies, cymbals, and noises, were of service towards dissipating melancholic thoughts;[63] the power of which we have accounted for in another place upon geometrical principles.[64] Hence also it more plainly appears, that the disorder was owing to natural causes; for otherwise how could the music of a harp drive it away? Counsel and prudence in a man was, in the Hebrew language, usually stiled the Spirit of God; and a person deprived of these qualities, was said to be troubled with an evil spirit, that is, to be mad.

I am not ignorant that the Jews, by a manner of expression familiar among them, are wont to describe diseases of this kind, to the power of evil angels, as ministers of God; and that even at this day, some very learned men defend the same notion. But for my part, if I may be allowed to declare my thoughts with freedom, I cannot think it right to have recourse to the divine wrath for diseases, which can be proved to have natural causes; unless it be expresly declared, that they were sent down directly from heaven. For if they fall on us in punishment of our sins, the intention of the supreme lawgiver would be frustrated, unless a sure rule was given, whereby his vengeance might be distinguished from common events; in as much as the innocent may be equal sharers in such calamities with the guilty. Moreover, it seems reasonable to believe, that evils inflicted by the omnipotent judge, must be either incurable, or curable by himself alone; that the connection of his power with his equity, may the more brightly shine forth. By such a criterion, are miraculous works distinguished from the operations of nature. For it would be impiety to suppose, that the almighty creator of heaven and earth intended, that his works should be performed in vain. Wherefore it is worthy of our observation, that great care is always taken in the sacred histories, to make the divine power in such cases, appear most manifest to all. Thus when the Lord had infected Miriam (or Mary) with a leprosy, for a sin committed by her, and consented, on the supplication of Moses, to make her whole; it was not done till seven days afterward.[65] Gehazi's leprosy remained in him and his progeny for ever.[66] King Azariah was smote with the leprosy, for not having demolished the high places; and he was a leper unto the day of his death.[67] Ananias and his wise were struck dead suddenly by the miraculous power of St. Peter.[68] Elymas the sorcerer, was struck blind for a season by St. Paul, for his frauds and wickedness.[69] Therefore since threats and plain indications of diseases, inflicted in an uncommon manner, are always manifestly declared; whensoever these are wanting, why may we not say, that the event was by no means supernatural? And I desire, once for all, that this sentiment may hold good with regard to several other calamities.

FOOTNOTES:

[62] See Samuel, or Kings, Book i. Chap. xvi.
[63] See Cels. Lib. iii. Cap. xviii.
[64] Mechanical Account of Poisons, Essay ii. Ed. 4.
[65] Numbers, Chap. xii. Verse 14.
[66] Kings, Book ii. (al. iv.) Chap. v. Verse 27.
[67] The same, Chap. xv. Verse 5.
[68] Acts, Chap. v.
[69] The same, Chap. xiii. Verse 11.

THE DISEASE OF KING JEHORAM

Of king Jehoram it is related, that, "for his wicked life, the Lord smote him in his bowels with an incurable disease, so that he voided his intestines daily for the space of two years, and then died of the violence of the[70] distemper." Two impious kings are recorded to have had the same end, Antiochus Epiphanes, and Agrippa; of whom it was said: Εἰς τί τὰ σπλάγχνα τοις □ σπλαγχνιζομένοις.[71]
Of what avail are bowels to those
who have no bowels?
Now this distemper seems to me to be no other than a severe dysentery. For in this the intestines are ulcerated, and blood flows from the eroded vessels, together with some excrement, which is always liquid, and slimy matter; and sometimes also some fleshy strings come away, so that the very intestines may seem to be ejected.

FOOTNOTES:

[70] Chronicles, Book ii. Chap. xxi. Verse 18.
[71] See the Notes of Grotius on this Place.

THE DISEASE OF KING HEZEKIAH

"When Hezekiah lay sick of a mortal disease, and the prophet Isaiah went and declared to him, by God's express command, that he should die and not recover; the Lord moved by his prayer, commanded Isaiah to return, and tell him, that he would cure him in three days. Whereupon Isaiah ordered a mass of figs to be taken, and laid it on the boil; whereby he recovered[72]."

Now to me it seems extremely probable, that this king's disease was a fever, which terminated in an abscess: For in cases of this kind, those things are always proper, which promote suppuration; especially digestive and resolving cataplasms; and dried figs are excellent for this intention. Thus, the Omnipotent, who could remove this distemper by his word alone, chose to do it by the effect of natural remedies. And here we have an useful lesson given us in adversities, not to neglect the use of those things, which the bountiful Creator has bestowed on us, and at the same time to add our fervent prayers, that he would be graciously pleased to prosper our endeavours.

FOOTNOTES:

[72] 2 Kings, Chap. xx.

THE DISEASE OF OLD AGE

Old-age itself is a disease, as the poet has properly expressed it[73]. Wherefore as I have frequently read with pleasure, the very elegant description of it, given by Solomon the wisest of kings; I think it will not be foreign to my design, to attempt an explanation and illustration thereof. For it contains some things not easy to be understood, because the eloquent preacher thought proper to express all the circumstances allegorically. But first I will lay the discourse itself before my readers, which runs thus.

"Remember thy Creator in the days of thy youth, before the evil times come, and the years draw nigh, in which, thou shalt say, I find no pleasure: before the sun, and the light, and the moon, and the stars be darkened, and the clouds return after rain; when the keepers of the house shall tremble, and the soldiers shall give way, and the diminished grinders shall cease; and those that look out thro' holes shall be darkened; and the doors shall be shut outwardly, with a low sound of the mill, and they shall rise up at the voice of the bird; and all the daughters of music shall be of no avail; also when they shall be afraid of high places, and stumblings in the way; and the almond tree shall flower, and the Cicadæ shall come together; and the appetite shall be lost, man departing to his eternal habitation, and the mourners going about in the street: before the silver chain be broken asunder, and the golden ewer be dashed in pieces; and the pitcher be broken at the fountain head; and the chariot be dashed in pieces at the pit; and the dust return to the earth, such as it had been; and the Spirit return to God, who gave it[74]."

The recital of evils (and infirmities) begins from the aberrations of the mind. The sun, says Solomon, and the light, and the moon, and the stars are darkened. Perceptions of the mind are less lively in old men; the ideas and images of things are confounded, and the memory decays: whence the intellectual faculties must necessarily lose their strength or power by degrees. Wisdom and understanding are frequently called light in the sacred scriptures;[75] and privation of reason, darkness and blindness.[76] Cicero likewise says very justly, that reason is as it were, the light and splendor of life.[77] Hence God is stiled the father of lights.[78] Thus the virtues of the mind decaying, may be compared to the luminaries of the world overcast. I am conscious that this exposition is contrary to that of a number of learned interpreters, who take this obscuration of the lights in the genuine sense of the words, and think that the failing of the sight is here to be understood. But I am surprized, how they happened not to take notice, that every thing in this discourse, even to the most minute circumstances, is expressed in words bearing a figurative sense. For whereas, in describing the infirmities of Old-age, the injuries of the operations of the mind, as the most grievous of all, were not to be pretermitted; so these could not be more clearly expressed, than by the obscuration of the cœlestial luminous bodies, which rule our orb, and cause the vicissitudes of times and seasons. Moreover it is particularly to be observed here, that the author mentions the defects of sight lower down, and most certainly he would have avoided repeating the same thing.

But he goes on, and adds, what well agrees with the foregoing explanation. The clouds return after rain. That is, cares and troubles crowd on each other, and daily oppress aged folks. As in moist climates, and those liable to storms, even when the clouds seem to be exhausted, others soon follow, and the rains become almost perpetual. And these inconveniencies are felt the more sensibly, in proportion to the debilitation of the powers of the mind, whereby they are rendered less able now, than formerly, either to bear, or get the better of their oppressions.

But from the mind our royal author now passes to the body. The keepers of the house, says he, shall tremble, and the soldiers shall give way, and the diminished grinders shall cease. The limbs, and firmest parts of the body, are damaged by age: the hands and knees grow weak, thro' the relaxation of the nerves. Hence those are rendered incapable of defending us against injuries, and of performing innumerable other good offices, for which they were originally intended; and these becoming unequal to the weight they were wont to sustain, lose their active suppleness, and fail in bending. Likewise the double teeth or grinders, either drop out, or rot away; so as now to be too few remaining to comminute solid food. In the translation of the Hebrew word, which I have here rendered by double teeth or grinders, I followed Arias Montanus, who, in my opinion, has translated it right. For it is in this passage used by the author in the plural number; who afterwards employs it in the singular, but in a quite different sense, when he treats of the sense of tasting; as I shall shew anon, when I come to that passage. For, that Solomon's intention in this place was, to describe those defects of the senses, which generally steal on old-age, I have not the least doubt.

Wherefore now proceeding to them, he begins by the sight. Those, says he, that look thro' holes shall be darkened. By which words it is manifest, that he points out the failing of the eyes, which most people, far advanced in years, feel by sad experience.

Next follows the taste, which he thus describes: The doors shall be shut outwardly, with a low sound of the mill. As old people, thro' diminution of appetite, open their mouths seldomer than formerly; so for want of teeth to comminute their food, they do it with less noise. Now this last inconvenience seems to be meant and expressed very elegantly by the words a low sound of the mill: for by the word mill, which in the Hebrew is used in the singular number, the grinding of the food may very well be meant; and this grinding, as it is not done by the assistance of the teeth, which they have lost, but by that of the gums, is performed with less noise.

Sleep is the sweet soother of our labours, and the restorer of our exhausted strength. But the loss of appetite, and disgust to our food, generally robs us of this comfort. Hence subjoining this evil of old-age to the foregoing, he says: he shall rise up at the voice of the bird; that is, the old man is awaked at the cock's first crowing. Wherefore his sleep is short and interrupted, tho' his weakness would require longer rest.

But he returns to the senses, among which he gives the third place to hearing; for receiving the benefits of which the Creator gave us the use of ears. Now this is frequently diminished, and sometimes entirely taken away in old-age; which the royal author seems to indicate in the following words: The daughters of music shall be of no avail. For thus he thought proper to express the ears, to which at this time of life, not only the pleasure of harmonious sounds is sought in vain; but, what is much more disagreeable, the words in conversation are not easily understood: whereby the enjoyment, and one of the greatest conveniencies of life, are gradually lost. Hence in the jewish history, Barzillai, at eighty years of age, complains that he could no longer hear the voice of the singing men and singing women.[79]

These defects of the organs of hearing, are immediately followed by those of the sense of feeling. Now the touch, as Cicero says, is uniformly spread over the whole body; that we may feel all strokes and appulses of things.[80] Wherefore this sense, besides its other uses, contributes vastly to the safety of the body, and the removal of many evils, to which it is perpetually exposed. And this the sagacious author seems to have principally in view, when he says: They shall be afraid of high places, and stumblings in the way. For as old folks are unsure of foot, even in a plain smooth way, by reason of the weakness of their limbs; so when they come to a rugged uneven road, thro' the dulness of this sense, they do not soon enough perceive the depressions or elevations of the ground whereby they run the hazard of stumbling and hurting their feet. Therefore they are not unjustly represented as being afraid.

The only one that remains of the senses is that of smelling, the diminution of which in old men, he describes with equal elegance and brevity in this manner: the almond tree shall flower. By which words he seems to mean, that old people, as if they lived in a perpetual winter, no longer perceive the agreeable odors exhaling from plants and flowers in the spring and summer seasons. That this tree flowers in winter, we learn from Pliny, who in treating of it says: The almond tree flowers the first of all trees, in the month of January.[81] I am not to learn, that these words are by most interpreters understood as relating to grey hairs, which being generally a sure token of old age, they would have us believe, are denoted by the white flowers of the almond tree. But then, who can imagine, that this wise author, after having indicated the defects of four of the senses, by clear and distinct marks, would designedly pass over the fifth in silence? Besides, white hairs are by no means to be esteemed a sure and indubitable token of old-age; since there are not a few to be found, who turn gray in the middle stage of life, before their bodily strength is any ways impaired. Moreover, what they say of the flowers of the almond tree, does not seem to agree with the things they mean by them: for they are not, strictly speaking, white, but of a purplish cast. Thus far concerning the senses: let us proceed to the remaining part.

The scrotal rupture is a disease common to persons far advanced in years; whether it be formed by the intestine or omentum slipping down into the scrotum, or proceed from a humor distending that part. In either case the part is tumefied. This pernicious disease the Preacher thought proper to compare to a grasshopper. The grasshopper, says he, shall be a burthen, Oneri erit locusta. For thus the Hebrew phrase is more literally translated, than by convenient cicadæ, the cicadæ shall come together, as the learned Castalio has rendered it. Indeed the Vulgate version has impinguabitur locusta, the grasshopper shall be fatted. The Septuagint Παχυνθῆ ἡ ἀχρίς. The grasshopper shall be fatted. The Arabic version, turned into Latin, pinguescet locusta, The grasshopper shall grow fat. But our English translation, The grasshopper shall be a burden. It is well known, that the Hebrew language is always modest, and that the sacred Writers, in expressing such things as belong to the genital members, abstain from indecent and obscene words, for fear of offending chaste ears, and therefore borrow similitudes from any other things at discretion. Which is particularly observable in the Canticum Canticorum, or Solomon's Song, written by our Author. Now the grasshopper, or locust, is an odd-shaped animal, made up chiefly of belly; and therefore, especially when full of eggs, may be said to bear some resemblance to a scrotum, swoln by a rupture.

These parts being thus affected, the wise author adds, the appetite shall be lost; wherein he does not attend so much to the appetite for victuals, as for those other things, which are sought after in the vigor of life. For as the author of the Art of Love has rightly said: Turpe senilis amor[82].

That old people are crushed to death by so great a heap of evils and infirmities, and depart to their eternal habitation, to the grief of their friends, can be no matter of wonder. But in the remaining part of the discourse we are admonished, that their miseries in this life are not confined within these bounds, but that sometimes there is still an accession of others. For loss of strength in old age does not terminate at the limbs, or extremities of the body; the spine of the back also loses considerably of its firmness, by the daily diminution of power in its muscles and ligaments: hence an old man can seldom stand upright, but stoops his body towards the earth, which is shortly to cover it. This part is likened to a silver chain, which is said to be broken asunder. For the vertebræ, of which it is composed, may be looked upon as the rings or links, and they give way outward by the bending of the body. Moreover the medulla oblongata, which passes through them, is of a silver or whitish colour.

These points, which we have hitherto handled, are very difficult of explanation. But the three inconveniencies, which close the discourse, are true ænigma's, and require an Oedipus to solve them. And as such an one, in my opinion, has not appeared hitherto, I will use my endeavours to do it. The golden ewer, says he, is dashed in pieces: the pitcher is broken at the fountain-head; and the chariot is dashed in pieces at the pit.

Old men are troubled with defluxions from the head to the nose, mouth and lungs; which are compared to water rushing out of a broken bottle or ewer. And the ewer is said to be of gold, to express the dignity of the head.

Nor does phlegm flow from the head alone; but other parts also pour forth their juices too abundantly or irregularly. For the serosities, which are secreted by the kidneys (whose cavity is even at this day named pelvis by Anatomists) runs into the bladder; which, by reason of the relaxation of its sphincter, as if the pitcher were broken at the fountain head, is not able to retain its contents a sufficient time. Hence an incontinence or dribbling of urine is continually troublesome.

Now, the evils hitherto enumerated lodge in particular parts; but the last calamity, both in this discourse, as well as in old people, is that the whole body is afflicted. The very course of the blood is interrupted; hence wretched man is seized with difficulty of breathing, apoplexies or lethargies. The heart also, the principle and fountain of life, sinks thro' want of its usual force, and the broken chariot falls into the pit. The ancients indeed did not know of the circulation of the blood; but they could not be ignorant, that it was moved thro' the body, that it cherished the viscera and members by its heat, and lastly, that it concreted and grew cold in death.

But nothing in this whole discourse is so much worthy of our serious attention as these words, with which he closes it. The dust returns to the earth, such as it had been; and the spirit returns to God, who gave it. For by these words his intention seems plainly to have been, to refute the ignorant notions of those, who thought that the soul perished with the body, and to assert its immortality.

FOOTNOTES:

[73] Terent. Phorm. Act. iv. Scen. i. v. 9.
[74] Ecclesiastes, Chap. xii. Verse 1-7. translated from Castalio's latin version.
[75] Job, Chap. xviii. Verse 5, 6, 7.
[76] Matthew, Chap. vi. Verse 23. John, Ep. i. Chap. ii. Verse 11.
[77] Academ. iv. 8.
[78] James, Epist. Chap. i. Verse 17.
[79] Samuel, (al. Kings) ii. Chap. xix. Verse 35.
[80] Nat. Deor. ii. 56.
[81] Lib. xvi. §. 42.
[82] Ovid. Amorum, lib. i. Eclog. ix. ver. 4.

THE DISEASE OF KING NEBUCHADNEZZAR

Those things, which are related of Nebuchadnezzar king of Babylon, appear so surprizing and contrary to nature, that some interpreters have imagined that he was really transformed into a beast. For "being driven from the company of men for seven years, his dwelling was with the beasts of the field, he fed on grass as oxen; his body was wetted with the dew of heaven; his hair and nails were grown like those of birds. At length at the end of that space of time, his understanding was restored to him, and he was established in his kingdom, and excellent majesty was added unto him. Now his crime was pride and the contempt of God[83]."

All these circumstances agree so perfectly well with hypochondriacal madness, that to me it appears evident, that Nebuchadnezzar was seized with this distemper, and under its influence ran wild into the fields: and that, fancying himself transformed into an ox, he fed on grass in the manner of cattle. For every sort of madness is, as I shall specify more particularly hereafter[84], a disease of a disturbed imagination; which this unhappy man laboured under full seven years. And thro' neglect of taking proper care of himself, his hair and nails grew to an excessive length; whereby the latter growing thicker and crooked, resembled the claws of birds. Now, the ancients called persons affected with this species of madness λυκανθρῶποι or κυνανθρῶποι; because they went abroad in the night, imitating wolves or dogs; particularly intent upon opening the sepulchres of the dead, and had their legs much ulcerated either by frequent falls, or the bites of[85] dogs. In like manner are the daughters of Proetus related to have been mad, who, as Virgil says,

—Implerunt falsis mugitibus agros.[86]

—With mimick'd mooings fill'd the fields.

For, as Servius observes, Juno possessed their minds with such a species of madness, that fancying themselves cows, they ran into the fields, bellowed often, and dreaded the plough. But these, according to Ovid, the physician Melampus,

—per carmen and herbas
Eripuit furiis.[87]
Snatch'd from the furies by his charms
and herbs.

Nor was this disorder unknown to the moderns; for Schenckius records a remarkable instance of it in a husbandman of Padua, who imagining that he was a wolf, attack'd, and even killed several persons in the fields; and when at length he was taken, he persevered in declaring himself a real wolf, and that the only difference consisted in the inversion of his skin and hair[88].

But it may be objected to our opinion, that this misfortune was foretold to the king, so that he might have prevented it by correcting his morals; and therefore it is not probable that it befel him in the course of nature. But we know, that those things, which God executes either thro' clemency or vengeance, are frequently performed by the assistance of natural causes. Thus having threatened Hezekiah with death, and being afterwards moved by his prayers, he restored him to life, and made use of figs laid on the tumor, as a medicine for his[89] disease. He ordered king Herod, upon account of his pride, to be devoured by worms[90]. And no body doubts but that the plague, which is generally attributed to the divine wrath, most commonly owes its origin to corrupted air.

FOOTNOTES:

[83] See Daniel, Chap. iv. and v.
[84] See Chap. ix. of Demoniacs.
[85] See Aetius, Lib. medecin. Lib. vi. and Paul. Ægineta, Lib. iii. Cap. xvi.
[86] Eclog. vi. 48.
[87] Metamorph. xv. 325.
[88] Observat. med. rar. de Lycanthrop. Obs. 1.
[89] See above Chap. v. p. 36.
[90] See below, Chap. xv.

THE PALSY

There are three paralytics recorded in the holy gospels to have been cured by Jesus Christ[91]. The case of one of these, which is the third, having some singularities in it, I shall relate the particulars of it in the words of St. John, "There is (says the Evangelist) at Jerusalem, by the sheep market, a pool, near which lay a great multitude of impotent folk, blind, halt, and withered, waiting for the moving of the water. For an angel went down at a certain season into the pool, and troubled the water: whosoever then first after the troubling of the water stepped in, was made whole of whatsoever disease he had. And a certain man was there, who had an infirmity thirty and eight years. When Jesus saw him lie, and knew that he had been now a long time in that case, he saith unto him, Wilt thou be made whole? The impotent man answered him, Sir, I have no man, when the water is troubled, to put me into the pool; but while I am coming, another steppeth down before me. Jesus saith unto him: Rise, take up thy bed, and walk. And immediately the man was made whole, and took up his bed, and walked."

This pool, or at least some other in its stead, is shewn to travellers even at this day by the friars who reside there.[92] But, what is much more to the purpose, Eusebius asserts that it actually existed in his time, and had two basons; both of which were filled every year by the rains, at a stated time; and the water of one of them was of a surprizing red colour:[93] which last phœnomenon he attributes, according to the vulgar opinion, to the sacrifices, which were formerly cleansed there. But I am clearly of opinion, that it was owing to a red earth or ocre, which is frequently found in baths, raised up from the bottom at certain times by the rains, and mixing with the water.

Commentators find more than one difficulty here. For first they enquire what sort of water this was; next why it could not exert its virtue without

53

being troubled; then what was the nature of this troubling; and lastly, concerning the angel they do not agree, who he was. Wherefore I will offer my opinion, in a concise manner, on these several points.

First then, mineral waters were in high esteem among the ancients for many diseases: they used them inwardly and outwardly, and recommended them for different distempers according to the nature of the mineral, with which they were impregnated. Thus in paralitic cases, Celsus recommends swimming or bathing in the natural sea or salt water, where it can conveniently be come at; where it cannot, even in water made salt by art.[94] And Pliny says, sulphureous water is useful for the nerves, aluminous for paralytics, or other relaxed habits of the body. He likewise adds; They use the mud of those fountains with advantage, especially if, when it is rubbed on, it be suffered to dry in the sun.[95] The same author relates strange things of some springs. In Boætia, says he, there are two springs, one of which retrieves the memory, the other destroys it.[96] In Macedonia two streams meet, one of them extremely wholsome to drink, the other mortal.[97] And other things of the same nature. To these may be added what Lucian, an eye-witness relates of the river Adonis in the country of the Byblii. The water of that river changes its colour once a year, and turning as red as blood, gives a purple tinge to the sea, into which it runs: and the cause of this phœnomenon he ascribes to its passing thro' mount Libanus, whose earth is red.[98] Nor is it foreign to the purpose to observe, that there are wonderful eruptions of water in some countries. In the province of Conaught in Ireland, there is a fountain of fresh water on the top of a high mountain, which imitates the tide, by sinking and overflowing twice a day.[99] A certain spring in Hungary in the county of Saros, is under the influence of the moon: since it is well known to increase with the moon's increase, to diminish with its decrease, and to run quite dry at the great change or new moon.[100] In fine, medicinal waters were not uncommon in Palestine, the accounts of which are collected by that great master of oriental literature, Hadrian Reland.[101]

Nevertheless those who contend for a miracle in this place, say that there are no baths known, which can cure all distempers; nor any that retain their virtue but one single month in the year: they likewise add, that it was the action of the angel troubling this water, that gave it its sanative qualities. Those who are of a different sentiment, enumerate a number of waters, which become salutary at certain times of the year, by being then charged with metallic salts; the mud of which being brought up from the bottom, has been serviceable in many diseases. Wherefore they say, it is not just to have recourse to a supernatural power for effects, which may be produced by the ordinary course of nature. But as far as I am able to judge of these contradictory opinions, a middle way between them seems to me to come nearest the truth.

For my notion of the matter is, that the water of this pool acquired its medicinal virtues from the mud settled at the bottom, which was charged with metallic salts, perhaps from sulphur, allum, or nitre. And whenever it happened that the water was troubled by any natural cause whatsoever, perhaps a subterraneous heat, or rains; these salts were raised up and mixed with it, and might well be beneficial to those, who went down into the pool, before the metallic particles subsided. Wherefore it is no wonder, that there lay, in the porches of this bath, which the evangelist says were five in number, a great multitude of impotent folk waiting for the moving of the water; and especially of such as laboured under those diseases, for which it was serviceable, as blindness, palsies, and decays. And it was very natural for every individual person to endeavour to get into it as soon as possible; for fear of being frustrated of their cure by the subsidence of the mud. Wherefore he who first stept in, experienced the virtue of the water.

The next circumstance to be observed is, that the fact here related, happened when there was a feast of the jews, that is, the pentecost. And we learn from Eusebius, that this method of curing prevailed but once in a year.[102] But it is well known that this feast was celebrated in the month of May or beginning of June: which is a very proper season for the virtues of medicinal waters. Upon which account the patients flock'd thither the more eagerly, that they might catch a medicine, which they could make use of but once a year.

Lastly, with relation to the angel, who is said to have troubled the water at a certain season; those who contend for a miracle, attribute the sanative quality of the pool to him. But we have already taken notice, that whenever any thing uncommon or surprizing happened, of which the jews could not investigate the cause, they were accustomed to say, it was done by the angel of the Lord. Yet it is possible, that God might have added this miraculous circumstance to natural effects, that this pool should be sanative, at one certain time of the year only, and that too, when the whole nation were assembled to celebrate their solemn festival; and to him only, who first went into it. The reason of which proceeding (if it be allowed to form a conjecture on the divine counsels) might perhaps have been, that God was pleased to testify by so manifest a sign, that he would not, as he had promised, entirely abandon his chosen people; before the coming of the Messias.

Wherefore upon the whole, this salutary virtue of the water, which might be medicinal by nature, seems to be so regulated by God, as at the same time to afford the jews a token of his presence. But the power of Christ, administered to this infirm man, a more noble remedy than that water, his evil-chasing[103] word. And this power was the more seasonable in this case, because the disease was of so many years standing, that it could not be removed by a natural remedy: whence his divine virtue shone forth the more brightly.

FOOTNOTES:

[91] See Matthew, Chap. viii. and ix., and John, Chap. v.

[92] See Cotovici Itinerarium Hierosolymitarum, Lib. ii. Cap. ii. and Maundrell's Journey from Aleppo to Jerusalem, 8vo. p. 107. Oxford 1714.

[93] Onomasticon urbium and locorum sacræ scripturæ, in voce Βηζαθά.

[94] Lib. iii. Cap. xxvii.

[95] Lib. xxxi. §. 32.

[96] Ib. §. ii.

[97] Ib. §. 19.

[98] De Dea Syria.

[99] Vid. Ortelii Theatrum orbis terrarum.

[100] Vid. Geo. Wernher. de admirandis Hungariæ aquis.

[101] Palæstina ex monument. vet. illustr. p. 300, andc.

[102] Loco citato.

[103] Αλεξικακον.

OF DEMONIACS

That the Dæmoniacs, δαιμονιζομένοι, mentioned in the gospels, laboured under a disease really natural, tho' of an obstinate and difficult kind, appears to me very probable from the accounts given of them. They were indeed affected various ways. For sometimes, they rent their garments, and ran about naked; striking terror into all those whom they met, and even wounding their own bodies; so very furious, that tho' bound with chains and fetters, they broke their bonds, and rambled in the most lonely places, and among the sepulchres of the dead. Sometimes also they cried out, that they were possessed by many devils, which they imagined could pass out of themselves into other bodies.[104] At other times, either they were worried, and made a hideous noise;[105] or were thrown on the ground, without being hurt, and the devil went out of them.[106]

These are all actions of madmen; but the dispute is, whether they were wrought by devils, or by the violence of the disease. Thus much is certain, that in those times it was a common opinion among the jews, that evil spirits frequently took possession of people, and tortured them in so surprizing a manner, as if they were agitated by furies. For in the whole catalogue of diseases, which afflict mankind, there is no other, that seems so much to surpass the force of nature, as this, in wretchedly tormenting the patient by fierce distractions of the mind, and excessively strong, tho' involuntary, motions of the body. But most certainly we find nothing sacred in all this, nothing but what may arise from a natural indisposition of body. And in order to place this my opinion in the stronger light, it may not be improper to give a short discourse on madness; not indeed on that species, which comes on in an acute fever, and goes off with it, which is called a phrenzy, and is always of short duration; but that other sort, which is rivetted in the body, and constitutes a chronical disease.

Wherefore all madness is a disease of an injured imagination, which derives its origin from the mind, having been too long a time fixed on any one object. Hence proceed uneasiness and anxieties of mind concerning the event. And by how much the things, whose images incessantly occur to the imagination, are of greater moment in life, the more violently they disturb the person; examples of which we see particularly in love and religion, wherein hope, fear, despair, and other contrary passions, succeeding each other by turns, drag the person different ways. That this is the case, will not be doubted by any one, who recollects, that a madman often has a good memory; manages his affairs, except when some vain ideas come across his mind, with tolerable prudence, nay sometimes with more than ordinary cunning; and that he ofttimes recovers the intire and permanent use of his reason, by a course of proper medicines. Therefore in this disorder the person is first over-whelmed by terrifying ideas, which are followed by wrath and fury, as attendants on anxiety: whence he threatens and attempts to do acts of the utmost cruelty to those who approach him, and thro' excess of anguish, frequently lays violent hands even on himself: then he grows again melancholic; and thus rage and dejection of spirits affect him alternately: moreover it is no uncommon thing to see a person under these circumstances, especially when the disease has taken deep root by length of time, seeking unfrequented and solitary places, in order to avoid the conversation of his fellow creatures,

Ipse suum cor edens, hominum vestigia vitans.[107]

Gnawing his heart, shunning the steps of men.

Now, people afflicted with this disorder, often live a long time. For all mad folks in general bear hunger, cold, and any other inclemency of the weather; in short, all bodily inconveniencies, with surprizing ease; as they enjoy a strength of constitution superior to what might be easily imagined. Likewise it frequently happens, that an epilepsy comes on madness of a long standing. For these diseases are nearly related; and in this case, we know by experience, that there remain not the least hopes of recovery. Lastly it is to be observed, that the patient is either frantic or melancholic, according as his habit of body is disposed to receive this or that injury.

But that the casting out of devils, is nothing more than the removal of madness, many do not believe, upon this account, that those things which happen to persons thus affected, seem to them impossible to be done by the force of nature. But certainly these gentlemen are too much strangers to physic, and have not sufficiently attended to phœnomena no less surprizing, which daily occur in other diseases. Do we not often see that violent affections of the mind are the cause of death? A sudden fright has destroyed many, and even excessive joy has been fatal. A dangerous distemper sometimes passes from one part of the body to another, in the twinkling of an eye. The venom thrown into the mass of blood by the bite of a mad dog, generally lies still a good while; and at the end of some weeks, sometimes months, exerting its strength, it produces symptoms not inferior to those, which are said to be produced by devils. What is more surprizing than some things which fall out in pregnancies? If a pregnant woman happens to have an eager desire for any thing, and is disappointed, she sometimes marks the fœtus with the figure or likeness of the object longed for, on this or that part of the body. And, what is still more, and approaches to a prodigy, upon the mother being terrified by a sudden injury done to any one part, that very part in the child suffers the same evil, and decays for want of nourishment. I know that the truth of stories of this kind, is called in doubt by some physicians; because they cannot conceive, how such things can happen. But many examples, of which I have been an eye-witness, have freed my mind of all scruples on this head. Now, the power of the imaginative faculty is so stupendous, that the mind is not less affected by false, than by true images, when daily subjected to them. This we find by experience in those women, who are called witches, who, being under the influence of such an error of the mind, frequently imagine that they not only converse with devils, but also have enter'd into compacts with them; and persist in these notions with such obstinacy, that, when they are brought to a trial, they confess themselves guilty of wickednesses, which they never perpetrated, though they know that they must suffer death for their confession. Moreover, every body knows how wonderfully the mind is disturbed in melancholies. One of them thinks his head is made of glass, and is afraid of stirring abroad, for fear of having it broken: another believes himself to be actually dead, and refuses food, because the dead ought not to eat. There are a thousand stories of this kind. I remember, a man of letters, with whom I was well acquainted, who positively asserted that he was big with child, and was vastly anxious for a happy delivery. I saw two others, who, when alone, fancied they heard the words of people whispering them in the ear. Nor is their case different, in my opinion, who persuade themselves that they see ghosts and hobgoblins. For deliriums are a kind of dreams of people awake; and the mind in both cases affects the body differently, according to the nature of its objects.

From what we have said, it manifestly appears, how many different ways the lessons of imagination, when they are confirm'd by long habit, are capable of affecting a man, and entirely ruining his whole frame. But every body knows, that the human mind is disturbed by nothing more than by fear; the cause of which is self-love ingrafted in all men. Whereas then, as Cicero very justly observes, there is no nation so savage, no man so rude, as not to have some notion of the gods;[108] it is no wonder, that men conscious of wicked deeds, should be struck with the fear of God, whose empire over all created things they acknowledged. For, as they attributed every good thing, every benefit of this life, to the gods; so they were of opinion, that evils and calamities were sent down by them in punishment of crimes. Now, idolatry, as I said above,[109] had its origin among the Chaldeans; and at first it consisted in the worship of the sun and moon, but afterwards it was extended to the adoration of dæmons.[110] But these were believed to be divine ministers; and that they were originally the souls of heroes and great men, who were worshipped for services done to mankind in general, or to their native country in particular. And this dæmoniac religion being propagated from the Chaldæans to the Phœnicians, then to the Egyptians, came afterwards to the Greeks, thence to the Romans, and in progress of time to the other nations.

But the jews, accustomed to ascribe every uncommon or wonderful work of nature to the agency of angels, as ministers of the supreme deity, could easily work up their minds to believe, that some dreadful diseases, which injured the mind and body together, the causes whereof they could not investigate, arose from the operation of evil angels. For we learn from Philo Judæus,[111] with whom Josephus also agrees in opinion, that they believed there were bad as well as good angels; that the good executed the commands of God on men, that they were irreprehensible and beneficent; but the bad execrable, and every way mischievous.[112] But a more illustrious example of this matter cannot be given, than in the narrative of Saul's disease,[113] of which I have already treated.[114] Nor were madness and the epilepsy the only diseases, which they imputed to devils. When Jesus had restored speech to the furious dumb man, he is said to have done it by casting out a devil.[115] And when he had cured another furious person, who was blind and dumb, the pharisees reproached him with casting out devils by beelzebub the prince of the devils.[116] In fine, Christ himself uses this common way of expression, on occasion of the woman which had a spirit of infirmity eighteen years, whom he freed from that infirmity; by saying, that satan had held her bound these eighteen years.[117]

And this custom of taking madmen for demoniacs, was not so peculiar to the jews, but that it prevailed in other nations also. Hence in Herodotus king Cleomenes is said to be driven into madness, not by any dæmon, but by a habit of drunkenness, which he had contracted among the Scythians, whereby he became frantic.[118] And whereas δαιμονᾷν signifies the same thing as δαιμόνιον ἔχειν, Xenophon uses this word for furere, to be raging mad or furious.[119] Moreover Aristophanes, intending to express a high degree of the same disease, employs the word κακοδαιμονᾷν, and calls the highest degree of madness, not μανίαν, but κακοδαιμονίαν[120]. Hence also, as Aretæus observes, this disease was called morbus sacer, or the sacred disease, because it was imagined that some dæmon had entered into the man.[121] Wherefore the physicians found it absolutely necessary to oppose this false notion with all their might. Because the people were generally persuaded, that diseases, which they believed to be caused by evil spirits, were to be expelled, not by medical skill, but by religious rites and ceremonies. Upon this account the prince of physicians Hippocrates, or at least some one of his scholars, wrote a very useful piece,[122] wherein he asserts that no diseases are inflicted on man, immediately, by any divine power; and that those persons ought to be accounted magicians and jugglers, who cover their ignorance with a veil of sanctity, by infusing such notions into the minds of the people.

But with regard to this power of the devils over human bodies, believed equally by the jews and other nations, I have already said, that the divinity ought not to be made a party concerned in imposing diseases, which may possibly have natural causes; unless it be expresly declared, that they were inflicted immediately by the hand of God.[123] For of all the diseases, with which miserable mortals are tormented, there are none so wonderful and dreadful to appearance, but may be the natural consequences of bodily indispositions. Wherefore God himself, if he thinks proper, can employ either natural causes, or the ministry of good angels, to inflict all sorts of diseases on mankind. And I hope nobody will believe, that the devils have had the power granted them of torturing men at their wanton pleasure. But to say more on this subject seems the less necessary; because two very learned divines of our nation have already treated it in a full and ample manner.[124]

Therefore in order to put an end to these demoniacal diseases, I will now briefly shew, how they are to be treated. And first of all, particular care should be taken, to keep the patient's mind employed in thoughts directly contrary to those, which possessed it before: for one set of ideas gives place to another, and by effecting this change, the mind is brought out of the state in which it was: a circumstance, to which the generality of physicians do not give sufficient attention. When this can be brought about, the disease is sometimes speedily cured. But when either the long standing of the distemper, or some other cause, renders this total change impracticable; at least the strength of the present set of ideas ought to be diminished and destroyed by all possible means. The vain fears of some are to be diverted, and their dismal thoughts dispelled. The daring ferocity of others is to be curb'd; for which end it is often necessary, to use hard words and threats. Likewise sudden frights, which may give the mind a different commotion, from that which before disturbed it, have been found to afford a temporary relief at least. The ancients prescribed some corrections, such as bindings and stripes.[125] And indeed it is sometimes necessary to bind those, who are too unruly; to prevent their doing mischief to themselves or others. But there is the less necessity for torments and stripes, because all mad men are of such a cowardly disposition; that even the most frantic and mischievous, after being once or twice tied, surrender at discretion, and thence forward refrain from committing any outrage, thro' fear of the punishment.

As to the medical part, the gross humors of the body are to be thinned, and the disorderly motion of the animal spirits is to be calmed. For which end blood-lettings, emetics, cathartics, blisters, and setons, also sometimes coolings of the head are to be employed. To these the fœtid gums are to be added, especially assa fœtida, myrrh, and galbanum. And camphire has been frequently found serviceable in excessive ferocity and want of sleep. But when the disease is accompanied by a fever, nothing is more proper than nitre, given in as large quantities as the stomach will bear. Lastly, the patient is to be kept to a slender diet, and compelled to use exercise. But in all evacuations, a certain degree of moderation ought to be used, lest the madness be changed into a contrary disease, which the ancients termed morbus cardaicus,[126] that is, an excessive weakness of body. In which case, the patient is so far exhausted, that medicines are of no avail; but the miserable dejected man drags the remains of life, alass! generally too long.

FOOTNOTES:

[104] See Matthew, Ch. viii. v. 28. Mark, Ch. v. v. 2. and Luke, Chap. viii. v. 27.

[105] Mark, Chap. i. v. 23-26.

[106] Luke, Chap. iv. v. 33-35.

[107] Cicero, Tuscul. Disp. Lib. iii. 26. who has turn'd into Latin this verse of Homer: "Ὃν θυμὸν κατέδων, πάτον ἀνθρώπων ἀλεείνων." Il. Z. v. 202.

[108] Tusc. quæst. Lib. i. 13.

[109] Cap. i. p. 5.

[110] See Sir Isaac Newton's Chronology, p. 160.

[111] Lib. de gigantibus.

[112] De bello judaico, Lib. vii. Cap. 6.

[113] See Samuel (or Kings) Book i. Chap. xvi.

[114] Chap. iii. page 28, andc.

[115] Matthew, Chap. ix. Verse 32.

[116] Ib. Chap. xii. Verse 22.

[117] Luke, Ch. xiii. v. 16.

[118] Lib. vi. Cap. 84.

[119] Memorabil. Lib. i.

[120] Vid. Plutum, Act. ii. Scen. 3. v. 38. and Act. ii. Scen. 5. v. 15.

[121] Διὰ τῆς δόξης δαιμονος ἐς τον ανθρωπον ἐισόδ□. De causis morbo diuturn. Lib. i. Cap. 4.

[122] De morbo sacro.

[123] Chap. iii. page 30.
[124] See the works of Jos. Mede 1677 fol. discourse vi. and enquiry into the meaning of demoniacs, andc.
[125] Vid. Celsus, Lib. iii. Cap. xviii.
[126] Idem, Lib. iii. Cap. xix.

OF LUNATICS

As some ancient physicians attributed the falling sickness to some divine power, so they ascribed madness to the influence of the moon. Yet the lunatic, σεληνιαζόμενος, whose disease is described in the gospels, was affected with the falling sickness.[127] Wherefore this patient (for there is but one of this kind expresly recorded there) was either mad and epileptic at the same time, which is not uncommon; or he laboured under a periodical epilepsy, returning with the changes of the moon, which is a very common case. For the account given of him is very short, that he ofttimes fell into the fire and oft into the water. Now in this distemper a person falls down suddenly, and lies for some time as dead; or by a general convulsion of his nerves, his body is agitated, with distorted eyes, and he foams at the mouth. But at length he recovers out of the fit, and has no more knowledge or remembrance of it, than if nothing had happened to him. Yet Jesus is said to have rebuked the devil, and he departed out of him, and the child was cured. That this child's case was epileptic, appears more manifestly from the account given of it by the evangelist, who was also a physician: for he says, that as soon as the spirit has seized the patient, he cries out, foams at the mouth, and is torn and worried by him.[128]

Now, as to these σεληνιαζόμενοι, who are subjoined to the demoniacs, as if their diseases were different, and whom Jesus is said to have cured;[129] they were either mad, or mad and epileptic together, which is not an uncommon case, as we have just now said. And as to devils, we have treated of them sufficiently. But with relation to the moon, there is not the least reason to doubt, but that the regular returns of the paroxysms at certain times of the month, gave occasion to men to believe, that this disease was lunar. For that planet has such a real influence on this disease, that it frequently happens to some patients, never to be seized with the fit but about the new and full moon; which seems to join its energy to those causes, that are adapted to produce this evil. But the manner of accounting for this I have delivered in another place; where I have plainly shewn that our atmosphere has its tides as well as the sea.[130]

And indeed the great Hippocrates has long since taught, that this disease is owing to natural causes, and consequently, by no means divine.[131] For altho' in his time, neither the inward parts of the animal body, nor the properties of the blood and humors, especially of the nervous fluid, were sufficiently known; yet by his great sagacity and experience, he has left us several useful observations, in relation both to the nature of the disease, and to its cure. For he has shewn, that it arises from too great a quantity of humors in the brain; and therefore that the best method of cure is to dry up, and lessen the quantity of this peccant matter; without having recourse to incantations and juggling tricks, so much in use in those days.

But when in succeeding ages, the use of medicines became more common, a great number of remedies for this dreadful disease were invented, some of which indeed were too filthy and shocking: such as drinking the warm blood of a gladiator just slain; eating human or horse's flesh, the testicles and penis of some animals, and other things of the same kind;[132] as if matters so repugnant to nature, could be contrary to such grievous defects of it. For so it often happens, that when a rational medicine is not to be found, any improper and rash one is attempted. But such experiments are to be abandoned to itinerant quacks, and credulous old women. Though even in our days our art is not sufficiently purged of this filth in these cases; seeing the dung of some birds, and the hoofs of quadrupeds are still ordered to be swallowed down by the sick. But whereas chemistry has furnished us with the means of extracting the salts, and other most active principles from bodies; to me it is matter of admiration, why physicians do not choose to order these principles to be taken pure into the body, rather than the coarse and fœculant substances, that contain them; which are always disagreeable, and sometimes hurtful also, to the stomach. But this most difficult distemper demands helps far superior to these; nor will any one method of cure answer in all cases, but the course must be altered according to the difference of constitution, andc. However, I will here propose those things, which have been found to be most generally serviceable.

Blood is to be taken away several times, according to the strength of the patient, in order to check its impetus. Vomits are to be administered now and then, but cathartics more frequently. It is particularly requisite to draw the redundant humor from the head, which is done by blisters; but better, by applying a caustic near the occiput, and making an issue, which is to be kept constantly running.

These remedies contribute indeed to weaken the paroxysms; but for removing the cause, when it can be done (for sometimes it cannot) other helps are requisite. For it is manifest, that the cause lies chiefly in the nervous fluid, commonly called animal spirits. But to investigate the manner how this fluid is affected in diseases of this kind, would, in my opinion, be a fruitless labour. However, as I have shewn on another occasion,[133] that it consists of very minute particles secreted from the blood in the brain, and receives and imprisons a considerable quantity of that elastic matter, universally diffused throughout all nature; it cannot be doubted, but that it may be so corrupted by some indisposition of the body or mind, as to become more or less improper for executing the functions of life, and perform all animal motions, not at the command of the will, but in a disorderly manner, and with a certain ungovernable impetuosity. Now the best remedies for correcting this depraved condition of the animal spirits, are chiefly those, which have the most powerful faculties of attenuating the humors, and throwing them out of the body by sweat. Of these the most excellent are the Root of wild Valerian, Russian Castor, the fœtid Gums, and Native Cinnabar, taken daily in pretty large quantities; with the interposition of cathartics at proper intervals, among which there is none better than the Tinctura sacra. I have long known by experience, that the celebrated Misleto of the Oak, is an useless weed. And indeed how can it be otherwise, since it has scarcely any taste or smell, and is entirely indebted to the religion of the Druids for its great character. Wherefore it is to be rank'd with those other frivolous things, which superstition has introduced into physick; unless a person can work himself up into a belief, that the golden sickle, with which it was cut down, the priest's snow-white garment, the sacrifice of white bulls, and other such trifling circumstances, are conducive towards a cure.[134]

FOOTNOTES:

[127] Matthew, Chap. xvii. v. 15 and 18.
[128] Luke, Chap. ix. v. 39, andc.
[129] Matthew, Chap. iv. v. 24.
[130] De morbo sacro.
[131] See influence of the sun and moon, Chap. i. and ii.
[132] See Celsus, Lib. iii. Cap. xxiii. and Cael. Aurelian, Lib. i. Cap. 4.
[133] Account of poisons, ed. 3. introduction.
[134] Plin. hist. nat. Lib. xvi. §. ult.

THE ISSUE OF BLOOD IN A WOMAN

Saint Matthew relates, that "Christ, by his word alone, cured a woman who had been diseased with an issue of blood for twelve[135] years."
And here arises a question, concerning the nature of this disease. But as the words in the Greek are γυνὴ ἁιμορρο□ σα, I am of opinion, that it was a flux of blood from the natural parts, which Hippocrates[136] calls ῥόον ἁιματώδη, and observes, that it is necessarily tedious. Wherefore having been exhausted by it for twelve years, may justly be said to be incurable by human art.

FOOTNOTES:

[135] Chap. ix. v. 20.
[136] De morb. Lib. i. Sect. 3.

WEAKNESS OF THE BACK, WITH A RIGIDITY OF THE BACK BONE

"There was a woman which had a spirit of infirmity eighteen years, and was so bowed together, that she could in no wise lift up herself, and Jesus laid his hands on her, and she was freed from her infirmity, and immediately made[137] strait."

This woman was συγκύπτ□σα, that is, stooping forward; being unable ἀνακύψαι, or to lift up her head. Now that spirit, according to the common way of speaking of the jews, was satan. For thus Christ himself, answering the ruler of the synagogue, who was angry that the woman had been cured on the sabbath day, says, that satan had held her bound these eighteen years. And exactly in the same sense saint Mark employs πνευμα ἄλαλον for a spirit, which obstructed the faculty of speech.[138]

This infirmity often befalls those, who have been very long afflicted with a disorder of the loins: whence the muscular fibres of that part become contracted and rigid. Wherefore it is very probable, that this tedious disease proceeded from that very cause, and was curable by the divine assistance only.

FOOTNOTES:

[137] Luke, Chap. xiii. v. 11, andc.
[138] Chap. ix. v. 17.

THE BLOODY SWEAT OF CHRIST

Saint Luke relates of Christ himself, that, "when he was in an agony by the fervency of his prayers, his sweat was like drops of blood falling down on the[139] ground."

This passage is generally understood, as if the Saviour of mankind had sweated real blood. But the text does not say so much. The sweat was only

ὡσεὶ θρόμβοι αἵματος, as it were, or like drops of blood; that is, the drops of sweat were so large, thick and viscid, that they trickled to the ground like drops of blood. Thus were the words understood by Justin Martyr, Theophylactus and Euthymius. And yet Galen has observed, that it sometimes happens, that the pores are so vastly dilated by a copious and fervid spirit; that even blood issues thro' them, and constitutes a bloody sweat.[140]

FOOTNOTES:

[139] Chap. xxii. v. 44.
[140] Lib. de utilitate respirationis.

THE DISEASE OF JUDAS

In the number of diseases, I rank the death of Judas, the wicked betrayer of Christ; of which I shall treat the more willingly, because very learned interpreters of the holy scriptures have run into different opinions concerning it. And about fifty years ago, two famous professors of history in the university of Leyden, Jacobus Gronovius and Jacobus Perizonius, handled this controversy in print with too much passion. For polite literature does not always polish its admirers.

The origin of the dispute was this. Perizonius had published Ælian's variæ historiæ, with his own notes and those of others; where taking occasion from what Ælian says of Poliager,[141] he diligently examines the signification of the verb ἀπάγχεσθαι, which saint Matthew[142] employs in relating the death of Judas; and insists that that word does not only mean strangling with a halter, but also sometimes excessive grief, by which a person is brought to the brink of death, and frequently even destroys himself. This criticism was taken amiss by Gronovius, who had already published a book de morte Judæ, wherein he had said that the wretch had voluntarily put an end to his life by a halter; wherefore he drew his pen, in order to refute his adversary's reasonings, and corroborate his own.

Moreover he quarrels with Perizonius about the phrase πρηνὴς γενόμενος, which he positively affirms ought to be understood not of a dying man, but solely of one actually dead, or of a dead body cast or tumbled down. For St. Matthew simply says ἀπήγξατο,[143] but St. Luke more fully, πρηνὴς γενόμενος ἐλάκησε μέσος, καὶ ἐξεχύθη πάντα τὰ σπλάγχνα αὐτ□,[144] that is, falling headlong, he burst asunder in the midst, and all his bowels gushed out. Wherefore, if the verb ἀπάγχεσθαι can bear no other signification than that strangling, which is performed by a halter, it is plain that the two evangelists do not agree together; unless we say with the learned Casaubon, that Judas hanged himself, but the rope broke, and he tumbled headlong down. But this does not explain the manner of his death; which saint Luke manifestly seems to have intended; but barely adds a circumstance of little moment, which happen'd after it, or at the very instant of it. Upon the whole it is certain, that by this word is not only meant suffocation by hanging, but also excessive grief, with which those who are violently overpowered, frequently compass their own death. For, as Ovid says: strangulat inclusus dolor. And indeed Perizonius has clearly proved this point by a number of examples, drawn from ancient authors.[145] Nor is it less to be doubted, but that the expression πρηνὴς γενόμενος, may be used for one, who voluntarily throws himself headlong down, as well as for one, who falls headlong by some accident: which he has amply demonstrated.

This controversy cost more than one dissertation. But after seriously considering the strength of the arguments produced by both parties; I am of opinion, that the words of saint Matthew may be reconciled with the account given by saint Luke from saint Peter's speech, in this manner. When that most unhappy traitor saw that Christ was condemned to death, he began to repent of his deed; and being thereupon wreck'd with grief and despair, or seized with the swimming in the head (which often happens in such cases) he fell headlong down some precipice; or, which is more probable, he designedly threw himself down, and his body chancing to pitch on some large stone or stump of a tree, his bowels burst forth, and he was killed. Wherefore Matthew declared his tortures of mind, which made him destroy himself; but Luke has clearly and properly determined the manner of his death. Thus this kind of death ought, with good reason, to find a place in the list of diseases, upon account of the real disorder of the mind.

FOOTNOTES:

[141] Lib. v. Cap. 8.
[142] Chap. xxvii. v. 5.
[143] Ibid.
[144] Acts, Chap. i. v. 18.
[145] Vid. dissert. de morte Judæ, and responsones duas ad Gronovium, Lugd. Bat. 1702 and 3.

THE DISEASE OF KING HEROD

The disease with which Herod Agrippa is said to have been smitten, by the just judgment of God, in punishment for his pride and of which he died, is remarkable. For he finished his miserable life σκωληκόβρωτος, that is, eaten by worms, as the sacred historian relates, in these words, "Upon a set day, Herod, arrayed in royal apparel, sat upon his throne, and made an oration unto them: and the people gave a shout, saying, it is the voice of a god, and not of a man. And immediately the angel of the Lord smote him, because he gave not God the glory; and he was eaten of worms, and gave up the[146] ghost." Josephus indeed, in his account of the fact, makes no mention of worms, but says that he was suddenly seized with violent gripings, and after being incessantly tortured with pains in his bowels for five days, he expired.[147] But saint Luke has informed us, that the worms, by which his bowels were eroded, were the cause of the gripes.

Now the greatest singularity in this king's disease is, that it was instantly inflicted on him from heaven (which he himself acknowledged according to Josephus[148]) otherwise as to verminose putrefaction in human bodies, we have several instances of it. For this very king's grandfather, Herod, surnamed the Great, is said to have labour'd under this disease a long time, till at length it threw him into a decay, of which he died.[149] Likewise Herodotus relates of Pheretima, the mother of Arcesilaus, king of Cyrene, that she was rotted alive by worms.[150] And it is recorded of the Roman emperor Galerius Maximianus, that this same loathsome disease not only eat away his genital members, but put an end to his life.[151] Wherefore it was impossible, but that some at least of the Greek physicians must have observed some cases of this kind. And accordingly Galen has proposed medicines for ulcers, σκώληκας ἔχοντα, that is, abounding with worms.[152]

For he says, in abscesses there are frequently found animals, ζῶα, very like those, which are engendered from corruption.[153] And Philoxenus in Aetius says, that in the humor of Atheroma's, he sometimes found animals, like gnats or little flies.[154] In fine, Paulus Aegineta teaches the method of getting rid of them.[155]

In so clear a case, it is needless to collect a greater number of authorities from the ancients, especially since several modern physicians have made the same observations. For Marcellus Donatus mentions a person of high rank, extremely fat, whose belly was eroded and mortified by little worms engendered in his skin, which was excessively distended by fat and humors; and these worms were not unlike those produced in old rotten cheese.[156] The learned Nicolaus Tulpius saw worms very like these, issuing with the urine out of the body of a very celebrated physician.[157] And the Ephemerides naturæ curiosorum, contain three remarkable cases of this kind. The first is that of a certain Frenchman, whose blood was so corrupted, that very minute animals came forth day and night with horrid tortures, thro' most of the outlets of the skin, as the eyes, nose, mouth, and bladder; and at length put an end to his miserable life.[158] In the second, black worms, not unlike scarabæi or beetles, came out of an abscess formed in the calf of the leg of a girl.[159] And in the third it is said, that very small white worms issued with the milk from the breasts of a woman in childbed.[160] Nor can I omit two similar cases, one of which is related by Poterius, the other by his commentator Frideric Hoffman. The former attended a countryman, for a tumor on his right knee, out of which, when opened, little live worms issued, which caused an intolerable pain in the part by their bitings. And the latter saw a tradesman, who had a hard tumor about the veins of the arms, which was very troublesome to him. This was opened by a surgeon several times without any benefit; until an ulcer was formed, out of which he took a great number of little black worms, armed with stings or prickles.[161]

Now these histories, wonderful as they seem, are not to be refused credit. For all nature is animated in a surprizing degree. The air which we breathe, the food which we eat; all fluids especially, are full of animalcula of very different kinds. Whence it is possible, that some of these, being received into our bodies, and conveyed into the minute passages of the softest parts, as into nests, may there grow, as worms do in the intestines, to their proper size. Hence by the obstruction of the smallest vessels, tumors arise; which being suppurated by heat, and bursting, pour forth their foul offspring in the shape of worms.

Wherefore I cannot agree with those interpreters, who imagine that Herod was consumed by, and died of the phthiriasis, or louzy disease. For σκώληξ is a different creature from φθείρ; this corrodes the surface of the skin, that the inner parts of the body. Nor can it admit of doubt, that saint Luke, who was a physician, well understood the meaning of both the words. And yet I know that the disease proceeding ὑπὸ τῶν φθειρων is by some learned men confounded with that caused ὑπὸ τῶν σκωλήκων; of the first of which Pherecides Syrius,[162] and Lucius Sylla,[163] are said to have died. Whereupon Kuhnius says,[164] I look upon the word σκωληκόβρωτος in saint Luke, and φθειρόβρωτος in Hesychius,[165] to be synonimous terms: and his reason is, because lice are worms.

THE END

FOOTNOTES:

[146] Acts, Chap. xii. v. 21-23.

[147] Antiq. jud. Lib. xix. Cap. viii. §. 2.

[148] Ubi supra.

[149] Josephus Ant. Jud. Lib. xvii. Cap. vi. (an. viii?) § 5. and De Bello Jud. Lib. i. Cap. xxxiii. § 5.

[150] Hist. Lib. 4. a fine Ζωσα ευλέων εξεζεσε.

[151] Sext. Aurel. Victor. Epitom. and Pompon. Laeti Rom. Hist. compend.

[152] De compos. Medic. per genera, L. iv. Cap. x.

[153] Lib. de tumorib. præter nat. Cap. iv.

[154] Lib. xv. Cap. vii.

[155] Lib. iv. Cap. xlii.

[156] De hist. medic. mirab. Lib. i. Cap. v.

[157] Observ. medic. Lib. ii. Cap. 1.

[158] Decur. 2. ann. 5. append. Artic. 38.

[159] Ibid. Artic. 52.

[160] Ibid. Artic. 109.

[161] Poterii opera cum annot. Frid. Hoffmanni edita, Francof. 1698. pag. 72.

[162] Ælian. var. hist. Lib. iv. Cap. 28.

[163] Plutarcho in ejus vita.

[164] Not. ad Ælianum.

[165] Lib. de vit. philos.

www.ingramcontent.com/pod-product-compliance
Lightning Source LLC
Chambersburg PA
CBHW070843180526
45168CB00002B/939